Understand Electronic Control Systems

Owen Bishop

Newnes

OXFORD BOSTON JOHANNESBURG MELBOURNE NEW DELHI SINGAPORE

Newnes
An imprint of Butterworth-Heinemann
Linacre House, Jordan Hill, Oxford OX2 8DP
225 Wildwood Avenue, Woburn, MA 01801-2041
A division of Reed Educational and Professional Publishing Ltd

 A member of the Reed Elsevier plc group

First published 2000

© Owen Bishop 2000

British Library Cataloguing in Publication Data
A catalogue record for this book is available from the British Library

Library of Congress Cataloguing in Publication Data
A catalogue record for this book is available from the Library of Congress

ISBN 0 7506 4601 2

Printed and bound in Great Britain by Biddles Ltd, *www.biddles.co.uk*

FOR EVERY TITLE THAT WE PUBLISH, BUTTERWORTH-HEINEMANN
WILL PAY FOR BTCV TO PLANT AND CARE FOR A TREE.

Contents

All photographs and drawings are by the author.

1
Introducing control systems

One of the simplest electrical control systems is a switch. In Fig. 1.1, a single-pole single-throw switch is used to control an electric heater. It turns it on or turns it off. This circuit is electrical rather than electronic, because it does not include any semiconductors. However, it can be linked to an electronic circuit, as will be shown later in this chapter.

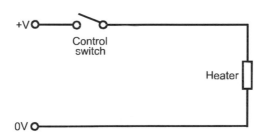

Figure 1.1 *The simplest control system. When the switch is closed, current flows through the heater. This converts electrical energy into heat energy.*

Note that the circuit works just as well whether its power supply is DC, as pictured in Fig. 1.1, or AC. This is because a heater works equally well whichever direction the current flows through it. The same circuit could be used to control a filament lamp or an electric bell. Some other devices, such as an electronic siren or a light-emitting diode (LED), must have the current flowing in the right direction, so that only a DC supply can be used. For controlling an electric motor, the direction of current might or might not make a difference, depending on the type of motor.

Fig.1.1 shows a simple type of switch that could be a lever switch, a slide switch or a rocker switch. There are many more specialised types of mechanical switch that could be used instead for different control applications. These include a tilt switch, a microswitch (which includes limit switches to detect the position of parts of a mechanism), a pressure pad (used in security systems), a vibration switch, a proximity switch, and a key switch on a keyboard.

In this chapter, we modify the basic circuit of Fig. 1.1 in various ways to illustrate the main features of electronic control systems. The first modification appears in Fig. 1.2. This is a truly electronic circuit that does the same thing as the circuit of Fig. 1.1, but relies on the principle of transistor action. The controlled device in this circuit is a filament

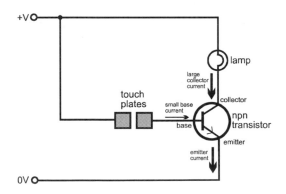

Figure 1.2 *A small current passing from plate to plate through a fingertip is amplified by the transistor to make a current large enough to illuminate the filament lamp.*

Control systems at work – 1

In an industrial chemical plant, a sliding shutter (1) allows powdered urea to fall from the hopper (2) into the reaction vessel below (3).

The shutter is moved by an *electro-pneumatic actuator*. This consists of a cylinder (4) in which slides a piston that is connected to the shutter. The piston is moved to close the shutter when compressed air is admitted to the left end of the cylinder, through an electrically operated valve (5). Another valve (6) admits air to the right end of the piston to open the shutter.

The travel of the piston/shutter assembly is detected by two *limit switches* (7 and 8). When the shutter is opening, it moves to the left until a projection on the piston/shutter assembly trips the switch (7). The closure of this switch produces a signal that is sent to the control circuit. On receiving this signal, the circuit closes the valve (6) and so prevents more air from entering the cylinder. The shutter is now fully open.

The reverse happens when the shutter is closing. In this case, valve 5 stays open until the closure of switch 8 informs the control circuit that the shutter is completely closed.

lamp. When a finger is placed so as to bridge the gap between the two metal touch plates, a small current flows through the fingertip and into the base of the transistor. The resulting collector current is large enough to light the filament lamp. This circuit illustrates the idea of the transistor switch. Many other kinds of device can be switched on or off by switching the transistor on or off. Other kinds of transistor can be used, such as JFETs and MOSFETs.

There is an additional stage of switching in Fig. 1.3. The transistor switches on a relay coil that switches on a heater. When the transistor is switched on by closing the control switch, the resulting current through the relay coil energises it. This creates a magnetic field in the coil, which pulls the relay contacts together. Current then flows through the heater. The advantage of using a relay is that it can switch a current larger than most transistors can safely pass. Another advantage, often more important, is that the relay can switch alternating current, as shown in Fig. 1.3. This means that we can use a low-voltage switching circuit to control a heater that takes its power from the mains supply.

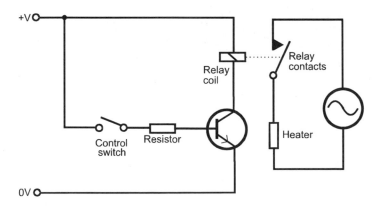

Figure 1.3 *A relay is often used to control a device that requires high voltage or current.*

Transistor action

A small current supplied to the base of a transistor causes a much larger current to flow from collector to emitter. In Fig. 1.2, the high resistance of the fingertip means that only a small base current flows. In Fig. 1.3, a resistor restricts the amount of current flowing to the base. It might be only a few milliamps, or even a few microamps.

Under the right operating conditions, the current that flows through the transistor from collector to emitter is proportional to the current flowing from the base to the emitter. Roughly speaking, it is about one hundred times greater. Therefore, the size of the collector current is usually several tens of milliamps or even several amps.

A direct or alternating current supply to high-current devices such as motors and heaters can also be switched by using a thyristor (Fig. 1.4) or a triac. In Fig. 1.5, a mains-rated lamp or motor is represented by the resistor R_L. Mains current flows through this when the triac is conducting. Since a triac consists essentially of two thyristors connected back to back, current can flow in either direction through it. This means that the triac may be made to conduct during both halves of the AC cycle. The pulse that triggers the triac is generated by the VR1/C1/diac network. It occurs at a definite point during each cycle, the time depending on the setting of VR1. The time can be varied from the beginning of a half cycle to the end of the half cycle. Once triggered, the triac conducts for the remainder of the half cycle. By adjusting the triggering point, we determine the proportion of the cycle for which current flows. This controls the amount of power delivered to the lamp or motor. In this way, the circuit acts as a brightness controller for a lamp or as a speed controller for a motor. Note that the previous circuits switch lamps and motors fully on or fully off. In this

Control by thyristor

As its symbol (Fig. 1.4) suggests, a thyristor is a switchable diode. Like a typical diode, it conducts only in one direction. If a voltage is applied across it in the correct direction, it still needs a short positive pulse at the gate electrode to make it conduct. After that, it conducts indefinitely until the voltage is removed or the current is reduced below a given 'holding' value.

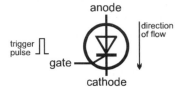

Figure 1.4 *A thyristor is a semiconductor device often known as a silicon controlled switch.*

Once it has stopped conducting, it does not conduct again until another pulse is applied to the gate.

circuit, VR1 is varied manually and the brightness or speed my be set to any level between fully on and fully off. The circuit is easily adaptable to an automatic control system.

In many industrial applications, the speed of a mains-powered motor is controlled by a low-voltage logic circuit, a computer or some other microelectronic system. There is the danger that a fault in the circuit linking the control circuit to the motor may break down and allow high voltages to pass to the low-voltage circuit. This will inevitably damage it, possibly beyond repair. There is another danger when driving motors and other highly inductive loads. Excessively large voltages may be induced in the control system when the load is switched off (see box). Even relatively small voltage pulses can cause errors in logic circuits. For example, a counter may be accidentally reset, or the

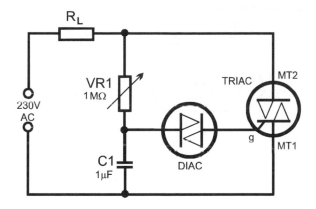

Figure 1.5 *This simple triac circuit gives variable control over the current passing through R_L.*

number stored may be changed to some other unpredictable number. One way to prevent this is to drive the heavy load through an *opto-isolator*. The opto-isolator (Fig. 1.6) consists of an LED and a phototransistor sealed in a lightproof capsule. When the LED is pulsed on by a current from the logic circuit, the phototransistor conducts for an instant and delivers a pulse to the triac. This switches it on for the remainder of the half cycle. Light passes across the opto-isolator but no current can flow from one side to the other. The device is constructed to that it can withstand a very high voltage difference, usually over 5000 V. Opto-isolators can also be used to prevent inductively produced pulses passing into the control circuits from the sensors in the system. We return to this topic in connection with programmable logic controllers in Chapter 5.

Summing up, an electronic device may be controlled by:

Direct mechanical switching.
A relay.
A transistor switch.
A thyristor or triac.

The control circuit may include one or more opto-isolators to prevent damage and interference.

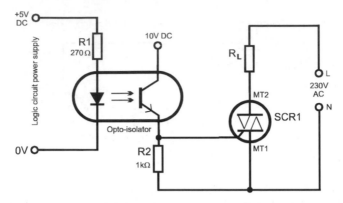

Figure 1.6 *An opto-isolator isolates the controlling circuit, with its low voltage and currents, from the actuator circuit, with its high voltage and current, and also its voltage spikes and other interference.*

Keeping up?

1 What are the advantages and disadvantages in using the four kinds of switches listed above?

2 State briefly what an opto-isolator does.

Electromagnetic interference (EMI)

One problem arising with thyristors and triacs is that triggering the device on and off generates high-frequency signals in the circuits to which it is connected. Such interference can spread to the mains supply lines to other circuits. Interference is worse if the thyristor is triggered on when the instantaneous mains voltage is near its most positive or most negative peaks. Then there is a relatively large increase or decrease in current, which generates the greatest interference. *A zero crossing detector* works by monitoring the instantaneous mains voltage and allowing triggering to occur only when the voltage is crossing from negative to positive or from positive to negative. At such times, there is only a small change of current as

Inductive loads

Any electronic device that has a coil is inductive, particularly if the coil is wound on an iron or ferrite core. Electric motors and solenoids are inductive devices commonly used in industrial plant for driving the machinery. It is a property of inductive devices that they act to prevent changes in the amount of current flowing through them. This is why we often use a coil or ferrite core (a *choke*) to smooth out peaks and ripples in the current being supplied to a circuit board.

When the current through a coil changes, the inductive device generates a voltage (called an *electromagnetic force*, or *emf*) to oppose the change. The size of the voltage depends on the *rate* of change of current. When a current is suddenly switched off, the rate of change of current is very high indeed. As a result, a very high voltage is generated, perhaps several hundred volts. Such voltages can easily damage sensitive components such as transistors. Voltage pulses generated in this way can also spread to other parts of the circuit, where they may upset the action of logical devices.

the thyristor is triggered and correspondingly less interference is generated. The CA3059 is a popular zero crossing IC (Fig. 1.7). It produces triggering pulses from pin 4 when the voltage at pin 13 is higher than that at pin 9. With S1 open as in the figure, the voltage at pin 13 is lower than that at pin 9, so no pulses are produced. However, if S1 is closed, there is a higher voltage at pin 13 and the triac is triggered at the beginning of every cycle. This is an elementary on-off circuit in which S1 could be replaced by a thermistor or light dependent resistor to make control of the load depend on light level or on temperature.

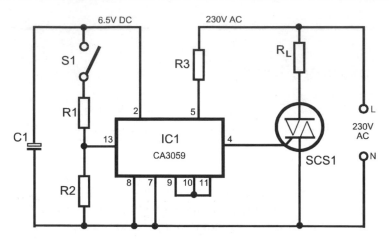

Figure 1.7 *IC1 is a zero-crossing detector. It triggers the triac at the very beginning of each cycle and so minimises the production of interference.*

Other ways of varying the voltage at pin 13 include using an opto-isolator to link the pin to the output of a logic circuit or computer. Another idea is to use a timer to generate a train of pulses fed to pin 13. By controlling the length of each pulse and the length of time between pulses it is possible to turn on the triac for a whole number of half-cycles (say 10 half-cycles, or 13 half-cycles) and off for another number of half-cycles. This provides variable amounts of power to the load, but in complete half-cycles so that minimum interference is generated. The technical name for a 'whole' number (with no fractions) is *integer,* so this technique is known as *integral cycle control.* It is also known as *burst fire control.*

Control systems

The action of the circuits of Figs. 1.1, 1.2, 1.3 and 1.6 is illustrated by Fig. 1.8. This is a *block diagram*, which shows the causative links between the different parts of the control systems. The first step in the switching on of the heater is a decision by the operator that the heater is to be switched on. The diagram shows that the operator interacts with the system at the control switch, causing the switch to close. The closing of the switch causes current to flow, either directly to the heater (Fig. 1.1) or to the base of the transistor (Fig. 1.2), or because of some other electronic link. Then the heater comes on. Each element in the system causes some action in the next element, the flow of action following the direction of the arrows. A block diagram does not tell us how the system is constructed but it is very useful for analysing how it works.

All control systems, even those that are highly automated, need an operator even if it is only to turn the system on and off. The operator is an essential element in the block diagram. Some control systems, such as those in spacecraft, are designed to run for days or even years without the intervention of an operator. Nevertheless, an operator is needed eventually to alter its action in some way. Conversely, there are many other systems that require the constant attention of a competent operator. Cars need sober drivers and airports need wide-awake traffic controllers. The qualities of the operator are important to the successful running of the control system. These qualities usually include skills and training.

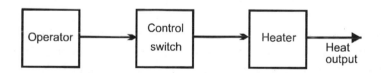

Figure 1.8 *A block diagram of a simple heating control system shows the relationships between causes and effects.*

Fig. 1.8 is incomplete. Few operators would close the switch, yet not look to see if the heater comes on. Fig. 1.9 includes a link back from the heater to the operator. The sight of the red glow of the element, or the sensation of warmth in the room satisfies the operator that the system is working properly. The operator takes no further action until the time comes for the heater to be turned off, perhaps when the room has reached a comfortable temperature. We say that the system in Fig. 1.8 is an *open* one, because the temperature of the heater has no effect

Sensors and actuators

Most control systems include at least one sensor and one actuator.

A **sensor** senses what is happening. Examples of sensors are switches, light detectors, thermistors (temperature sensor), and tilt-switches.

An **actuator** makes something happen. Examples of actuators are motors, solenoids, and lamps.

on the system. In Fig. 1.9 there is a *closed* system. The temperature of the heater affects an earlier stage in the system (the operator). The operator decides 'too hot' or 'too cold' and switches the heater off or on accordingly. A well-trained operator is continuously monitoring the output of the system and turning the heater on or off, as necessary.

Most control systems have one or more closed loops; primarily to ensure that what is intended to happen really does happen. In the example above, the operator completes the loop, using the switch to keep the room at a comfortable temperature.

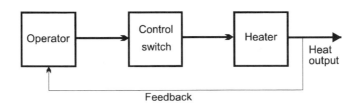

Figure 1.9 *Feedback from output to operator closes the control loop of the system of Fig. 1.8.*

Keeping up?

3 List some examples of control systems in which (a) an operator is seldom required to take action, and (b) the operator must be continually monitoring the output of the system.

4 List examples of (a) open, and (b) closed control systems.

Feedback

In Fig.1.9, a small fraction of the heat from the heater reaches the operator, providing information about whether the heater is on or off. We call this *feedback*. A similar example of feedback is found in a *thermostat* system. This kind of system is commonly found in cookers, domestic hot water systems, and refrigerators. As the name thermo-*stat* implies, its function is to maintain a static or constant temperature. Fig. 1.10 represents the main elements of such a system. We will

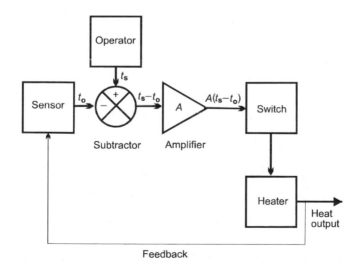

Figure 1.10 *A block diagram of a thermostat system shows how feedback from output to sensor is used to keep the room temperature constant.*

describe its properties and action as if the thermostat were controlling the air temperature of a room. The same principles apply if it is controlling the water temperature in a hot water tank. The same principles apply too, but in a reverse sense, if we are describing the action of a refrigerator.

In Fig. 1.10, the feedback of heat from the heater is detected by a heat sensor. The sensor circuit produces an output voltage, t_0, which is proportional to the air temperature. As well as being able to turn the system on and off by a switch (not shown), the operator is able to set the required temperature. By turning a knob or moving a lever, the operator is also able to produce a variable voltage t_s. This voltage is proportional to the temperature that the operator wants the air to come to. This temperature is called the *set point*. The operator can change the set point at any time, and immediately the circuit acts to bring the room to the new set temperature.

Thus, there are two important voltages:

t_0, proportional to the present room temperature, and
t_s, proportional to the set point, the desired room temperature.

The aim is to make t_0 equal to t_s, that is, to make the room temperature equal to the set point temperature.

These two voltages are fed to a *subtractor*. As indicated by the '+' and '−' signs in its symbol, the output voltage of the subtractor is equal to the difference between the two input voltages, t_s and t_0. To increase the sensitivity of the circuit, the signal from the subtractor is then fed to an amplifier that has a gain of A. The resulting voltage signal is:

$$A (t_s - t_0)$$

This is known as the *error signal*. It is a measure of the amount by which the actual temperature of the air differs from the set point. The error signal is sent to a switch circuit that controls the supply of power to the heater. In its simplest form, this could be a transistor switch and

Negative and positive feedback

The feedback of Fig. 1.10 and most other control systems is *negative.* If the sensor detects that the air temperature is too *high,* the heater is switched *off.* If the sensor detects that the air temperature is too *low,* the heater is switched *on.* In this way, the air around the heater is kept at a more-or-less constant temperature. In Fig. 1.10, feedback operates in the negative sense because t_0 is9 proportional to temperature and is *subtracted* from t_s.

Negative feedback operates so as to oppose any changes in the system. Negative feedback brings *stability.*

Positive feedback has the opposite effect. A small change in the output of the system is fed back so as to cause the output to deviate even further. The output swings strongly in one direction or the other, and often the system locks up with the output at its maximum positive or negative level. In other systems, the output may swing strongly and continuously between positive or negative. For this reason, positive feedback is often used in oscillators. For this reason too, it has very few applications, if any, in control systems.

relay as shown in Fig. 1.3. If the room temperature is exactly equal to the set point temperature, the error signal is equal to zero. If the room temperature is higher than the set point ($t_0 > t_s$), the error signal is negative. If the room temperature is lower than the set point ($t_0 < t_s$), the error signal is positive. The switching circuitry is such that only a positive signal turns on the switch to send current to the heater.

Practical thermostat

There are many possible ways of turning the block diagram of Fig. 1.10 into a practical thermostat system, and Fig. 1.11 is one of the simpler solutions. It is based on an operational amplifier (op amp, for short) wired as a differential amplifier. The action of a differential amplifier is to amplify the difference between two input voltages. The equation relating output to input is:

$$v_{OUT} = R_F \times (t_s - t_0)/R_A$$

Comparing this with the equation for Fig. 1.10, it is clear that A, the amplification or gain of the amplifier, is equal to R_F/R_A.

The sensor R1 in Fig. 1.11 is a thermistor. The resistance of this increases with increasing temperature. As connected in Fig. 1.11, the thermistor causes the voltage t_0 at the junction of R1 and R2 to increase as temperature increases. The value of R2 is generally chosen so that t_0 is close to 0 V when the temperature is at the centre of its operating range. The relationship between temperature and t_0 is not linear, in fact it is mathematically rather complicated, but is reasonably linear over a limited range.

The variable resistor VR1 is used to provide the set point voltage, t_s. The operator adjusts VR1 to produce a value of t_s that sets the

---

Thermistor

This is a bar, disc or bead of material that changes its resistance with temperature. The thermistors used in the circuits in this book have a negative temperature coefficient, which means that their resistance decreases as temperature increases. Thermistors are widely used in circuits that control temperature.

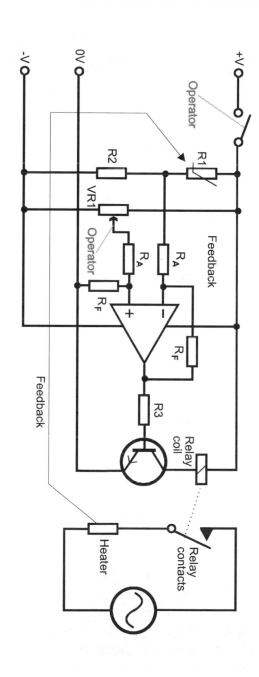

Figure 1.11 *A practical thermostat system uses an op amp comparator to detect whether the temperature is higher than or lower than the set point. Its output is fed to a transistor switch which controls a relay to switch the heater on or off.*

thermostat to the required temperature. With inputs t_0 and t_s applied to its negative and positive input terminals, the op amp produces an output signal v_{OUT}, as expressed by the equation above. This is fed to the base of the switching transistor. The heater is turned on whenever this signal is greater than 0 V. This happens whenever t_s is greater than t_0. The effect of this is that the air is heated until its temperature reaches the set point. This statement is not quite correct because the transistor does not switch on until its base-emitter voltage exceeds 0.6 V, so there is an *offset error* in the system. This can be corrected by calibrating the scale of VR1 to compensate for the offset, and also to take account of the non-linearity of the sensor.

In most thermostats, the sensor is situated at some distance from the heater. In a domestic central-heating system, for example, there are many heaters (the convectors, often inaccurately called 'radiators'), one or two in each of the main rooms. The sensor is often situated in the entrance hall or some other central room and is generally not positioned directly above one of the convectors. When the heaters are switched on, they heat the air immediately in contact with them. Convection currents carry this warmed air around the room and the air temperature rises gradually as this warm air mixes with the cooler air in other parts of the room. It may take 5 or 10 minutes for air temperature near the thermal sensor to rise sufficiently to cause the heaters to be switched off. The result of this is that feedback from heater to sensor is not immediate. There is a delay or *lag* in the system.

If the lag is too great, parts of the room may become too cold before the heater is switched on again. Conversely, parts of the room nearest to the heater may become uncomfortably hot before it is switched off.

However, a small amount of lag is advantageous. The air temperature is not exactly constant but fluctuates by two or three degrees about the set point. In many applications, a small fluctuation of this sort is acceptable. It is preferable to an excessively rapid response that has the relay contacts making and breaking so frequently that the relay soon wears out.

Mechanical thermostat

For comparison with electronic systems, Fig. 1.12 illustrates a purely mechanical system. The sensor is a *bimetallic strip,* which consists of two layers of different metals, each with different thermal expansivities. As shown in the figure the strip is more-or-less straight when cool, but curls as the strip becomes warmer. In practice, the strip is usually rolled into a helix that becomes more tightly wound as the strip becomes warmer. This allows a longer (and so more sensitive) strip to be contained in a reasonably small volume.

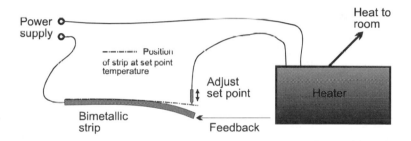

Figure 1.12 *This mechanical thermostat employs a bimetallic strip, which curls away from the contact when the temperature exceeds the set point.*

Electrically, the thermostat has the circuit shown in Fig. 1.1. At temperatures above the set point, the strip curls away from the adjustable set-point contact and the heater is off. As the system cools the strip straightens and eventually touches the set-point contact. This completes the circuit and the heater is turned on.

Feedback is negative because the system turns the heater *on* when the temperature *falls*. This ensures that the system is stable.

Keeping up?

5 Given that the gain A of the amplifier is 5, the set point t_S is 2.6 V, and the output t_0 is 1.9 V, what is the error signal?

6 Describe examples of feedback in a system consisting of a car and its driver.

Hysteresis

In the thermostat system described above, the temperature of the room fluctuates a few degrees above and below the set point. Putting it another way, the heater is not switched off until the temperature at a given part of the room has risen a few degrees above the set point. Conversely, it is not switched on until the temperature has fallen below the set point. This kind of action, in which switching takes place at different temperatures, according to whether the system is warming up or cooling down, is known as *hysteresis*.

A limited amount of hysteresis is an advantage in some systems. As mentioned earlier, it minimises contact wear in a system switched by relays. In a robot in which motors are used to control the position of an arm, it is wasteful of power to have the motors continuously adjusting the arm to a position that is being determined with an unnecessarily high degree of precision. The system hunts indefinitely about the set point when it would be better for it to get sufficiently close to the set point and then switch off.

Fig. 1.13 illustrates a system that ignores the error signal when this is less than a certain amount. The result is to create a dead zone, in which the arm is not necessarily exactly at the set point but is near enough to it to perform its function properly. The point at which the arm stops moving depends on the direction from which it approaches the set point.

Hysteresis is unavoidable in the case of the thermostat. It is inherent in the relationship between the parts of the system. In a typical thermostat

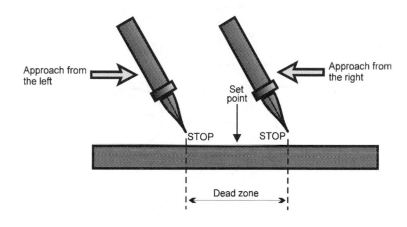

Figure 1.13 *To avoid the robot arm continuously hunting from side to side of the set point, movement of the arm stops when it is sufficiently close to the set point. This creates a dead zone in which the arm is stationary.*

system, the amount of hysteresis depends partly on the distance between the heater and the sensor and on the ease with which part of the heat is fed back from the heater to the sensor.

Often the amount of hysteresis in a heating system is not critical, but it can be minimised by placing the sensor directly on top of the heater. In fan ovens and computer enclosures we reduce though not completely eliminate hysteresis. In a hot water bath, a stirrer performs a similar task. In a central heating system, a circulating pump is used to reduce hysteresis.

Hysteresis can produce unacceptable results in some systems. Fig. 1.14 shows a system intended for providing at least a minimum light level on a surface in varying daylight. Here the feedback is light which, unlike heat, 'flows' instantaneously. This usually results in two unacceptable behaviours:

- If daylight is more or less steady at the level that *just* triggers the lamp to switch on, the lamp flickers and the relay (if used)

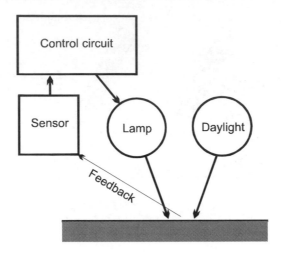

Figure 1.14 *With direct connection between the sensor and the control circuit, it is possible for undesirable effects to occur.*

chatters continuously. The flickering lamp is irritating and leads to early failure of the lamp and undue wear on the relay.

- If too much light from the lamp reaches the sensor, negative feedback switches the lamp off, at which point negative feedback switches the lamp on again. The system oscillates indefinitely, independently of the level of daylight.

Obviously, a simple system such as Fig. 1.14 is fraught with problems. Some of these can be avoided, or their effects lessened by introducing hysteresis. There is no hysteresis inherent in the system, so it is provided electronically. Fig. 1.15 shows a Schmitt trigger circuit added to the system as part of the loop. A Schmitt trigger can be built from a pair of transistors plus a few resistors, or it can be made from op amps or logic gates. Whatever units are used to build it, its action is the same and is illustrated in Fig. 1.16. Suppose that the sensor circuit has a voltage output (V_{IN} in Fig. 1.15) which is proportional to the amount of light falling on the sensor. The dashed line of Fig. 1.16 shows how V_{IN} varies when the light increases from darkness to full

Schmitt trigger

This is defined as a bistable circuit triggered by a rising or falling input. The triggering voltage for a rising input is higher (higher threshold) than the triggering voltage for a falling input (lower threshold). The difference between these voltages is the hysteresis.

brightness, then decreases back to darkness. The output of the trigger is V_{OUT}, represented by the solid line in Fig. 1.16. This is high (about 5 V) to start with when the sensor is in darkness. It could be fed to a transistor switch in the control circuit to turn the lamp on. Alternatively, it could be fed to a relay for the same purpose.

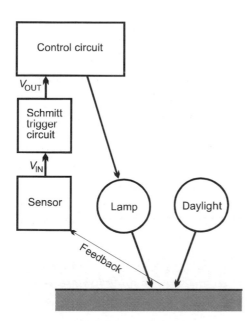

Figure 1.15 *Introducing a Schmitt trigger circuit into the control loop avoids the undesirable effects inherent in Fig. 1.14.*

The Schmitt trigger circuit has hysteresis. Its output stays high until the input exceeds a given value known as the *upper threshold*. In the case of the trigger in Fig. 1.16, this threshold is at 2.5 V. As V_{IN} increases above 2.5 V, V_{OUT} rapidly falls to 0 V. The lamp is turned off. Even though the light level, and therefore V_{IN}, rises slowly, the fall in V_{OUT} is always rapid. This 'snap' action is characteristic of Schmitt triggers, and in this case prevents the lamp from flickering on and off as light reaches the triggering level.

Now comes the behaviour that gives the Schmitt trigger its other useful property. Once the circuit has been triggered by V_{IN} rising above the upper threshold, it stays in that state until V_{IN} falls below the *lower threshold*. Fig.1.16 shows that the lower threshold is about 1.4 V. as V_{IN} falls below 1.4 V, V_{OUT} falls almost instantly to 0 V and the lamp is extinguished.

The difference between these two thresholds is 1.1 V in this case, which is the hysteresis of the circuit. Switching levels and thus

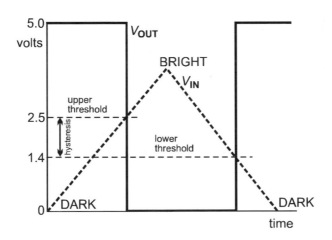

Figure 1.16 *The graphs of the changes of input and output voltages of a Schmitt trigger circuit, show that the hysteresis is the difference between the lower and upper threshold voltages.*

hysteresis can be set by choice of resistor values in the trigger circuit. In practical terms, this circuit prevents the chattering of contacts and the flickering of the lamp described earlier. Once light level has risen above the upper threshold and the lamp has gone off, a small reduction in light level is insufficient to return the circuit to its ON state. The snap action provides a firm response, so preventing chattering. The same applies after the light level has fallen below the lower threshold. If the light from the lamp does not produce too great an effect on V_{IN}, this circuit is immune to the effects of negative feedback. The lamp comes on at dusk, stays on until dawn, and then switches off until dusk.

Keeping up?

7 Design a system using two water-level sensors and an electrically operated water tap which will keep a water tank reasonably full, yet prevent it from overflowing. Define the hysteresis in your system.

Disturbances

The systems have been described above as if they are operating under steady conditions. In the thermostat, the room temperature is brought to a steady value and remains fairly close to this from then on. The only loss of heat from the system is through the walls, floor, and ceiling of the room. The amount of heat from the heater is adjusted by the thermostat to exactly equal this heat loss. In real life, such a steady state rarely lasts for long. Eventually, someone leaves the door open on a cold day and the room cools rapidly. This is described as a *disturbance*.

If the heater is powerful enough, it may be possible for the system to recover after a while. However, a strong breeze blowing in through the open door may break off the feedback between heater and sensor, and the system will go out of control permanently. The only solution is to shut the door, or to install a door-shutting spring!

Disturbances of the kind just described may turn out to be catastrophic but, if we recognise in advance that they may occur, the system can be designed to accommodate them. In the previous example, a more powerful heater or door spring may be installed. Alternatively, we may design for *proportional control*, as described in Chapter 2.

Disturbances can often be catered for by using feedforward. In a system that operates by feedback, the disturbance (leaving the door open) has to occur before it can be corrected.

Feedforward

Two examples from a wood turning factory show how feedforward may be used to avert disaster. Wood turning produces large quantities of sawdust. This creates a fire risk and, if mixed with air in the correct proportions, is explosive. In the factory, a system of pipes draws air from the working area of each sawdust-producing machine. This air is conducted to the outside of the factory, where the dust is filtered out and collected. The air is expelled by large fans. There is the danger of the dust catching fire and possibly exploding. Fire sensors in the main pipeline are connected to water-valves *further along* the pipe to spray water into the air when fire is detected.

As a second precaution, the amount of sawdust in the air in the pipeline is continuously monitored. If it approaches the proportions of an explosive mixture, the plant is automatically shut down.

Fig. 1.17 shows a system in which the disturbance is anticipated and corrected *before* it occurs. In a steel rolling mill, a slab of hot steel is passed between a series of rollers, which gradually reduce it to a strip of the required thickness. In each case, the vertical position of the low roller is fixed and there is a mechanism to raise or lower the upper roller. As the slab emerges from between the first pair of rollers (Rollers A) its thickness is measured by Sensor A. This is a radiation detector with a source of radiation (either a radioactive source or an X-ray generator) mounted below the path of the strip. The amount of

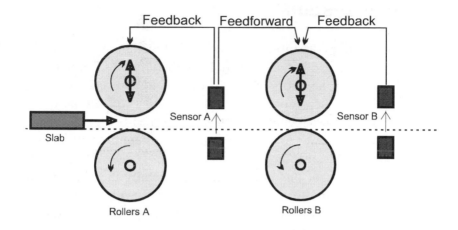

Figure 1.17 *Information obtained by sensors may be fed both backward and forward along a production line.*

radiation penetrating the strip is proportional to its thickness. There is the usual system of feedback from Sensor A to Rollers A to raise or lower the upper roller to obtain a strip of the required thickness.

The mill may also include *feedforward*. Because of wear in the rollers and the effects of temperature, Rollers A may not produce the required thickness. As well as the feedback going to Rollers A, a feedforward signal goes to Rollers B. There the upper roller is moved up or down to compensate for irregularities in the action of Rollers A. The timing of the compensatory action is such that it occurs after a slight delay, just as the appropriate section of the strip is passing through the rollers.

Feedforward is often a rough correction, intended to avoid the worst effects of the disturbance. In this example, it can not take into account errors caused by wear in Rollers B. So the system includes feedback from Sensor B, which, at the same time as it regulates the reduction in strip thickness produced by Rollers B, corrects the errors introduced by feedforward from Sensor A.

Kinds of system

Most of the systems described in this chapter function by switching a heater, fan, lamp, or other actuator fully on or fully off. The possible range of values of the error signal is divided into two regions. If the value of the signal lies in one region, the actuator is turned on. If the signal lies in the other region, the actuator is turned off. The actuator either runs at full power or not at all. Such a system is known as *an on-off system*, or sometimes as a *bang-bang system*. Very many control systems are of this type, which is generally the cheapest to build, to operate, and to maintain. They contrast with *proportional control systems* in which the power applied to the actuator is proportional to the error signal.

Proportional systems are described in more detail in the next chapter, but the principle of integral cycle control is an example in this chapter in which a proportional *effect* is produced by what is essentially an on-off system. The heater or other actuator is always switched fully on or fully off but the ratio between the number of on cycles and the number of off cycles can be controlled by suitable circuitry. In this way, the average power developed by the actuator over a short period of time can be controlled.

As well as categorising systems by the way the actuator is energised, we may also classify them according to the kind of control they provide. As its name implies, a thermostat is intended to hold the temperature of a room, an oven or a tank of fluid at a fixed temperature, the set point. It holds this temperature in spite of any changes in temperature in the surroundings. It is usually possible to vary the set point but in most instances the set point is a temperature that is to be held as steadily as possible and is rarely altered. For example, we might raise the set point of a central heating system in winter to allow for increased heat loss through the walls. However, we would not expect to alter it more often than, say, once a month.

Because of its action described in the previous paragraph, we say that a thermostat is a temperature *regulator*. There are many other examples of control systems that are regulators, including speed regulators in CD players, voltage regulators in electronic circuits, and chlorinating regulators in swimming pools.

Some control systems are designed with the intention that the set point will be altered frequently, perhaps continuously. Such a system is known as a *controller.* The power steering system of a car is an everyday example. In a chemical processing plant, for example, the temperature may need to be held at a series of fixed levels for varying periods. The set point is altered to a sequence of different levels during processing. The controller circuit is basically similar to a thermostat but there may be features to help the system to change rapidly from one temperature to the next. These features could include heaters more powerful than those normally used in thermostats, coolers to reduce temperature quickly, and fans or stirrers to bring the system to its new temperature more rapidly. The set point may be varied manually or automatically. A system such as this, in which the set point varies and the system follows these variations, is known as a *servo system* (see Chapter 3).

Test yourself

Answers to these multiple-choice questions are given in the supplement.

1 An opto-isolator:

<blockquote>
A turns off the power in an emergency.
B provides feedback.
C makes the control circuit more stable.
D prevents electrical damage to the control circuit.
</blockquote>

2 A device that performs an action when the control circuit switches it on is called:

<blockquote>
A a triac.
B an actuator.
C a Schmitt trigger.
D a sensor.
</blockquote>

3 A triac is triggered on by:

    A a positive pulse at its gate.
    B EMI.
    C a negative pulse.
    D an opto-isolator.

4 When a triac is switched on, it conducts:

    A in both directions.
    B only for the remainder of the half-cycle.
    C for a whole cycle.
    D indefinitely.

5 A relay is:

    A an electrically operated switch.
    B a semiconductor device.
    C a timing device.
    D a safety device.

6 The disadvantage of triac circuits is that they:

    A generate EMI.
    B conduct current in only one direction.
    C are low-powered devices.
    D easily become overheated.

7 A Schmitt trigger switches on and off at different voltages because of:

    A EMI.
    B hysteria.
    C positive feedback.
    D hysteresis.

8 If the voltage input to a Schmitt trigger is increasing, the circuit switches on:

    A at the lower threshold.
    B half way between the lower and upper thresholds.
    C at 0.6 V.
    D at the upper threshold.

9 In a servo control system, the output:

> A closely follows changes in the input.
> B remains very stable.
> C falls as the input rises.
> D slowly oscillates.

10 If a control system is designed to respond to changes in input conditions, we say that it has:

> A negative feedback.
> B feedforward.
> C hysteresis.
> D a servomechanism.

2
Control strategies

The simplest strategy, and one that is most often used, is the on-off or bang-bang technique described in the previous chapter. The actuator is simply turned fully on or fully off while bringing the system to the set point. In proportional control and the other systems described in this chapter, the actuator is turned on to a variable degree or off.

Proportional control

Fig. 2.1 shows the simple proportional system. Like most control systems, it relies on negative feedback. The error signal is the difference between the system output and the set point. The error

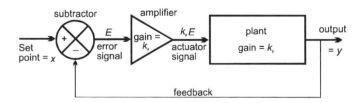

Figure 2.1 *In a proportional control system, the output is fed back and subtracted from the input to produce the error signal. This is amplified to produce a signal that drives the actuator.*

signal is usually amplified and a larger signal (proportional to the error signal) is sent to the actuator. There may be further gain in the actuator itself and this together with the amplifier gain makes up the *open loop gain* of the system.

An unexpected feature of this system is that the error never reduces to zero. There is always a difference between the set point and the output, or *product variable*. This difference is known as the *offset*. Fig. 2.2 shows a simple proportional control system set up on a simulator. The result of running the system is displayed on an 'oscilloscope' in Fig. 2.3. The lower curve is a plot of the input to the system which is a square wave with amplitude 1 V. Think of this as the set point, which is changing between 0 V and 1 V once a second. The upper curve plots the output from the system. This is also a 1 Hz square wave but, whereas the set point (input) changes from 0 V to 1 V, the output changes from 0 V to only 0.952 V. The offset is 4.8%. This is the result obtained when the gain of the loop is set at $K = 20$.

Studying control systems

The action of electronic circuits may be simulated by computers. The software has models, which simulate the action of electronic components such as resistors, transistors, and logic gates. As well as this, some simulators have models of the circuit blocks commonly used in building electronic control systems. These include gain blocks (amplifying circuits), adders and subtractors, integrating blocks and so on. A simulated control system is built up by 'wiring' these blocks together. Each block represents an electronic subsystem that may possibly be very complicated electronically. In the simulation, the blocks have 'ideal' properties including infinite input impedance and zero output impedance.

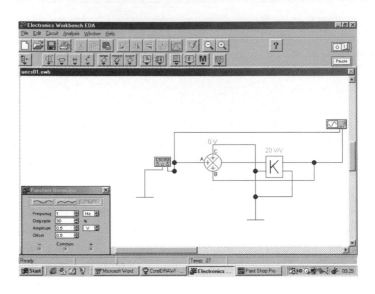

Figure 2.2 *This is the equivalent of the proportional control system of Fig. 2.1 set up on a simulator. Input comes from a square-wave generator, regularly changing the set point from 0 V to 1 V. The adder block is set with a gain of +1 for input (A) and a gain of –1 for feedback (B). The gains of the amplifier and plant are combined into a single gain block (K), here set to ×20. The input and output signals are read on the 'oscilloscope', with the results displayed in Fig. 2.3. The third input to the adder block (C) is connected to 0 V as it is not used in this system.*

The offset varies with the gain of the loop. The simulator is re-run with a different value for *K*, and the offset is found by measurements on the display of the 'oscilloscope'. When the gain is reduced to ×5, for example, offset increases to 16.7%. When the gain is reduced to ×2, offset increases to 33.3%. Reducing the gain increases the offset.

By contrast, increasing the gain reduces the offset. If the gain is increased to ×400, for example, the offset decreases to 0.25%. Output is then very close to the set point. It can be shown that the output is very close to the set point when the gain is large. It follows that the error signal is almost zero.

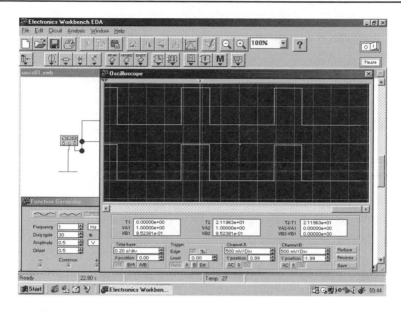

Figure 2.3 *The enlarged 'oscilloscope' shows the input (lower curve) alternating between 0 V and 1 V. When K = 20, the output alternates between 0 V and 0.952 V.*

For accurate proportional control, we need to make the offset as small as possible. This means making the open loop gain as large as possible. Usually, the gain in the actuator depends only on its physical characteristics and can not be altered. High open loop gain is obtained mainly by increasing amplifier gain. This gives the system an accurate response, but may introduce problems. There is also the possibility of the system becoming unstable and breaking into oscillations.

It is easy to show that there is always an offset in a proportional control loop (see Fig. 2.2), but it is harder to understand how it occurs. The point to note is that the error signal at any instant is amplified and is included in the output signal. Then it is inverted and fed back, then amplified *again*. It circulates the loop indefinitely, being inverted each time round the loop. Being alternately positive and negative, the circulating error values tend to cancel each other out. However, the cancelling effect is not exact, because of the amplification. The result is a permanent bias or *offset* in the values.

Proportional + integral control

The error offset described in the previous section can be reduced to zero by adding another signal input to the system, as in Fig. 2.4. An operator feeds in the signal while watching output until it reaches the required set point. The operator has to repeat this adjustment every time the set point is changed.

Fig. 2.4 illustrates the adjustment signal being added in with the input and feedback signals, but it can be added anywhere in the loop. Often it is added to the actuator signal between the amplifier and the actuator in the plant. Although many systems use manual adjustment of this kind, it is generally preferable for the correction to be applied automatically.

When the operator applies an adjustment manually, the signal is applied for as long as is necessary to bring the error to zero. In addition, the operator varies the size of the adjustment signal,

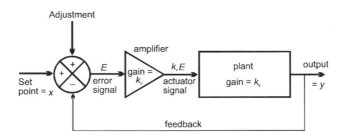

Figure 2.4 *The error signal inherent in proportional control may be reduced to zero by feeding in an adjustment signal, here controlled manually.*

Errors in proportional control

The error signal is the difference between the input or set point (*x*) and the output (*y*):

$$E = x - y$$

This signal is multiplied by the gain k_a of the amplifier, to provide the actuator signal k_aE. This signal is multiplied by the gain, if any, occurring in the plant, giving the output $y = k_ak_pE$.

To simplify the equations, put $k_ak_p = K$.

K is the *open loop gain* of the system and $y = KE$.

Substituting the value for *E* given in the first equation:

$$y = K(x - y)$$

From this we obtain:

$$y = Kx/(1+K)$$

This equation shows that if *K* is large, then *y* approaches the value of *x*. Output approaches the set point. If this is so, then the offset approaches zero.

depending on the error apparent at that time. Without thinking about it in such terms, the operator is applying a time-dependent and error-dependent correction. Fig. 2.5 shows how time and error are combined into a single function, the *time integral* of the offset. Before integration, the error is multiplied by a factor k_i to produce the most effective correcting action.

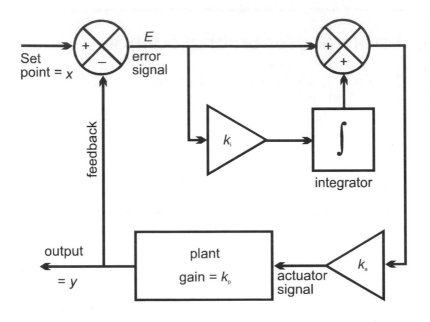

Figure 2.5 *In P+I control, an integrating block continuously produces a signal that is proportional to the error signal.*

Fig. 2.6 plots the input and output of a proportional plus integral (P+I) system set up as in Fig. 2.5, with K = 5. With proportional only control, the error is 16.7%, so the output rises almost instantly to 0.833 V and stays there. With P+I control, the graph in Fig. 2.6 reveals that output rises immediately to 0.862 V, a slightly higher level because the integral action has already begun. After this rapid initial rise, output *continues* to rise and reaches 0.973 V after 0.5 s. Error is then only 2.7%, a great reduction of error compared with the P only system. If the input signal were to stay at 1 V for longer than 0.5 s, the error would be reduced even further. We could adjust the gain of the integrator to produce a more rapid corrective response and bring the output closer to 1 V. The reverse action occurs when the input falls to 0 V. There is a rapid initial fall in output, followed by a continued fall toward zero as integral action continues to operate.

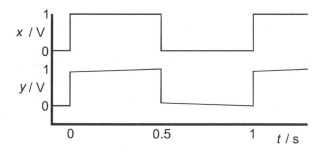

Figure 2.6 *When there is a change of set point (top curve), the output (lower curve) shows a rapid change due to proportional action. There is still an offset, as in Fig. 2.3, but this is gradually corrected by integral action.*

Keeping up?

1 How may we minimise the offset error of a proportional control system?

2 In what ways is proportional control improved by adding integral control to the system?

P + I + D control

Integral control eventually reduces the offset error to zero, but there are occasions when a more rapid correction is required. For example, opening the door of a furnace may cause a sudden drop in temperature. Once the door is closed again, P+I control will eventually bring it back to the set point. In the meantime, the serious reduction of temperature may produce critical stresses in the objects being heated.

If the furnace were being operated manually by an experienced person, the operator would take steps to bring the temperature back to its set point more rapidly. Perhaps the air draught and fuel flow could be increased temporarily until temperature is near the set point. When controlling a furnace manually, the operator is continually watching out for any rapid changes in the error. Unfortunately, integral action can not rapidly compensate for *rapid* changes. In an automatic system, we need to monitor the error signal in a way that detects rapid changes. In mathematical terms, the quantity of special interest is the derivative, dE/dt. A proportional system that incorporates derivative control as well as integral control is known as a three-term system, often abbreviated to P+I+D control (Fig. 2.7).

Fig. 2.8 plots the input and output signal of a typical three-term system. The plots of the input and output signals are identical for all practical purposes. For example, with the values used in the simulation, a rise of input from 0 V to 1.000 V results in a rise of output from 0 V to 1.007 V, an error of 0.7%. The rapid rise of input (high dV/dt) has brought about a steep rise of output. During the next

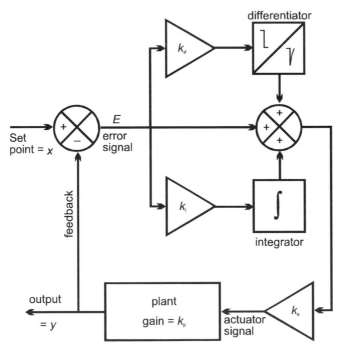

Figure 2.7 *Adding a differentiator (or derivative) block to the P + I system (Fig. 2.5) allows the system to repond to rapid changes in input. This is known as a P + I + D system.*

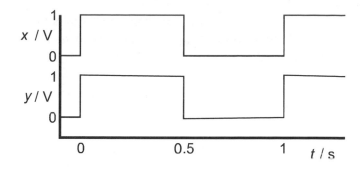

Figure 2.8 *With P+I+D control, the output (lower curve) can be made to follow the input (upper curve) almost exactly.*

0.5 s the output falls slightly as the corrective action of the integral block takes effect. Output falls to 1.004 V in that time, reducing the error to 0.04% in 0.5 s. If the high input pulse had been longer there would have been sufficient time for output to have become equal to input. At the end of the pulse, the input falls to 0 V, and the output almost instantly follows this change.

With appropriate 'tuning' of the gains of the amplifiers, three-term control gives a highly precise response to changes in the set point. Control is probably better than is required for the majority of industrial processes. Consequently, the additional expense of installing three-term control may not bring any real benefit. It can also cause problems. In the case of step increases or decreases in the input signal, it is possible that the rate of change, dV/dt may be so great as to cause wild swings in output. These may be damaging to the plant. In other situations, where there is electrical or electronic noise in the system, the noise itself is differentiated. This enhances the effect of the noise and the system is continuously being thrown out of balance by irregularly sized and irregularly timed swings of output. Noise may be the random voltage or current fluctuations that are present in any electronic system, or it may be a characteristic of the process itself. An example of the first type of noise is electronic noise in the output of a sensor, or in the output from its amplifier. An example of the second type of noise is irregular motion of parts of mechanical actuators due to friction or to badly machined or worn bearings. Another example is the random convection flows of liquid being heated in a vat, giving rise to random variations in temperature readings.

Keeping up?

3 What are the advantages of derivative control?

4 What are the possible problems of a system that includes derivative control?

System characteristics

The responses plotted in Figs. 2.3, 2.6 and 2.8 are those of an ideal system, behaving according to the mathematical equations governing proportional, integral and derivative control. The system responds instantly to changes in the input and produces an output that may be calculated and plotted, as in the figures. Time comes into the calculations because the input varies with time, perhaps very regularly as in the given examples. In addition, integral control involves integration over time, and derivative control depends on rates of change in time.

The output of a system may vary with time for other reasons, even if it a simple bang-bang system. Systems do not necessarily respond *immediately* to a change in input, or to a disturbance in their operating conditions. For example, the rotational inertia of a massive turntable tends to keep it rotating at constant speed, even when the torque of the driving motor varies. Inertia is advantageous in the case of a hard disk drive. Inertia is a nuisance in some other mechanisms. For instance, in turning a massive structure such as a crane from one position to another, we have to accelerate it rapidly to begin with, so that it will reach its desired position in a reasonably short time. Then we have to decelerate it, starting before it has reached its intended position, to bring it to rest pointing in the required direction.

The storage of energy in the rotating object creates a *first order lag* in the system. This modifies or even dominates the response of the system to changes in input. Also, it may create problems in controlling the system.

First order lags occur in other control systems that involve some kind of storage. An oven and its contents store heat energy supplied by the heater. Heat is lost from the oven at a rate depending on the difference between its temperature and that of the surroundings. Heat loss depends on how effectively the oven is insulated. One result of this is that, although the heater is supplying heat at a constant rate, the temperature of the oven does not increase in a linear way. The higher its temperature, the more slowly the temperature rises. Temperature increase is exponential.

The electronic equivalent of the heating oven is a charging capacitor. If the capacitor is being charged from a constant voltage source, the voltage between its plates rises exponentially. We say that the system has a *time constant*. When there is a change in input, the system moves 63% of the way toward the new state in one time constant. It reaches its new state in approximately five time constants. In both cases, the lag adds further complications to the relationship between input and output.

Integral action may be intrinsic in a system that includes a storage element. Fig. 2.9 illustrates a system intended to maintain a constant water level in a storage tank. The tank itself is the integrator since the amount of water in it, and hence the water level, depends on the sum of the various inflows and outflows. Part of the outflow is under the control of an operator as water is pumped away for use in, say, irrigation of an area of land. Part of the outflow may be the result of evaporation from the tank. This is dependent on weather conditions so it is irregular and unpredictable. Also dependent on weather (assuming an open-topped tank) is inflow due to rainfall. The tank integrates all these quantities in time and the resulting level is read by a sensor. The output from the sensor is compared with the set point and error signal

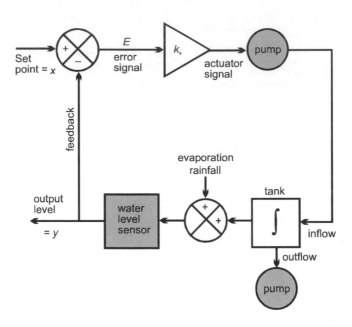

Figure 2.9 *A system that includes an element such a water tank may have the equivalent of integral action. It is then able to operate with zero offset when only proportional control is provided.*

to control a pump, which adds water to the tank as the level falls. In the figure, we are assuming that maintaining the level is always a matter of topping up. However, we could extend the system to excess water by making the actuator signal switch on the outflow pump when the water level is too high. The important point about this system is that although it is a simple proportional system, it acts like a P+I system because of the storage function of the tank. As a result, there is no error offset in its output.

In plant driven by motors, a flywheel helps keep the machine running at constant speed. It irons out any small change in speed that might be caused by minor changes in load, by fluctuations in the power supply or by other minor disturbances. This is because the rotational inertia of the flywheel acts as a store of energy. A large flywheel has large inertia and this is able to absorb excess energy without undue increase in speed. Conversely, it can make up for a transient excess of load on the system without a significant decrease in speed. Even without a flywheel as such, any mechanical system with rotating parts (including the rotor of the motor) has inertia. The system can not be brought to a halt instantly without applying active deceleration. This is a problem in position control of, say, a robot arm or a spoiler on an aeroplane wing. If the deceleration is insufficient, the part overshoots its intended position. By the time deceleration has begun to move it back toward the intended position, the inertia of the system may be sufficient to overshoot in the reverse direction. After several overshoots in alternate directions, with decreasing amplitude, the part settles at the intended position. Overshooting is often difficult to avoid and sometimes is acceptable in the interests of keeping the system simple. There are some situations in which it can never be allowed to occur. For example, when a spoiler is retracted, overshooting could permanently distort the wing surface.

Second order lag is a feature of certain types of control system, especially those concerned with controlling position and liquid levels. A closed-loop system often shows the features of a second order lag. Here the electronic equivalent is a capacitor and inductor wired in parallel. This gives a resonant circuit, which will oscillate strongly if its input varies at the resonant frequency. Often, the output oscillates with greater amplitude than the input. Once oscillation has begun, it may take some time to die out. For example, the level of liquid in a tank surges regularly, perhaps between dangerously low and dangerously high levels. A system such as this may be very difficult to control unless it contains elements to dampen the oscillations.

Damping is an important feature of systems with second order lags. Otherwise, oscillations may increase until the system gets out of control and is possibly damaged. As mentioned above, a small amount of damping (underdamping), sufficient to gradually damp out oscillations with a small amount of overshoot, may be acceptable in certain applications. Overdamping, in which the output changes to its new value without overshoot, may be essential in certain circumstances. The system must approach the new value slowly; otherwise, overshoot will be difficult or impossible to avoid. It follows that overdamping makes the system slow to respond.

Another feature that complicates control systems is *transit delay*. For example, in Fig. 2.10, cartons are being filled with washing powder at one point in the system. The cartons are weighed at a point further along the production line. The rate at which powder is fed into the cartons is determined by the control system and there is feedback from the weigher to adjust the amount supplied. However, there is a delay due to the transit time during which a given carton passes from the feeding stage to the weighing stage. This depends on the distance between the stages and the speed with which the cartons are moved. This delay limits the accuracy with which the amount of powder in each carton can be controlled.

Transit delay occurs in computer-controlled systems because a computer does not monitor all stages of the system continuously. It samples the readings from the various sensors at intervals. Consequently, there is a delay between one sampling and the next before there can be any change in the output of the system.

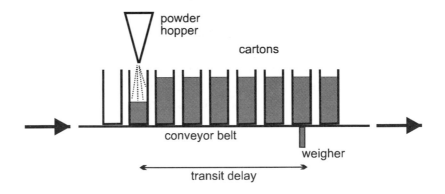

Figure 2.10 *If two connected processes are separated in time, as on a production line, this introduces transit delay into the system.*

Keeping up?

5 Explain the meaning of first order lag, second order lag, damping, and transit delay.

An alternative strategy

Proportional control, with or without integral and derivative corrections, is suited to bringing a system to a set point, or possibly to shift it from one set point to another. However, the nearer the system approaches the set point, the slower the rate of change. An underdamped system may hover on either side of the set point for an appreciable time. There are applications for which this type of response is unsuitable. An example is the control of the setting of an aileron, spoiler, or other flight surface of an aeroplane. The spoiler must be moved from one angular position to another as quickly as possible. Moving it as quickly as possible to a new position means that the spoiler must be accelerated at the maximum possible rate that will not cause it or its bearings to be damaged. Moreover, if it is being moved to lie flush with the wing surface, it must stop *immediately* it reaches that position. Damage would occur if it was allowed to dip below the wing surface. Usually, the actuator is a motor with reduction gearing to rotate the spoiler. A sensor on the bearings measures the angular position of the spoiler and provides feedback to the controller.

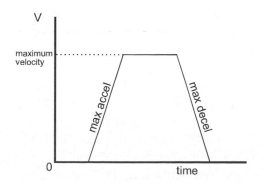

Figure 2.11 *A spoiler is accelerated and decelerated at the maximum safe rate, so as to bring it to its new position in the minimum time.*

Fig. 2.11 is a graph of the angular velocity against time. The slope of the curve shows the acceleration of the spoiler. As the spoiler moves in one direction, it is accelerated at the maximum safe rate (the upward slope of the curve) until it reaches the maximum safe angular velocity (the horizontal section of the curve). It moves at this velocity for a while and is then decelerated (downward slope) at the maximum safe rate. This is probably, but not necessarily, equal to the maximum acceleration rate. It continues decelerating at this rate until it comes to rest. The timing of the three sections of the curve are such that it comes to rest at its required position. These timings are calculated by the control circuit and depend on the initial position and the required final position.

In most such systems, the power actually applied to the motor must take into account any resistance to the motion of the actuator. In the case of flight surfaces, air pressure generates a variable force depending on the attitude and motion of the aeroplane. At each instant the desired position of the spoiler is calculated and then compared with its actual position. The power to the motor is varied accordingly.

A similar situation may occur in the control of other kinds of process. In the chemical process illustrated on p. 3, the rate of addition of urea must be kept under continuous control. This is because the reaction between urea and formaldehyde is *exothermic*. It generates heat. If the urea is added too slowly, the reaction is slow and much of the small amount of heat generated is lost to the surroundings. The process takes too long to complete. On the other hand, if the urea is added too quickly, the mixture may overheat and even explode. There is an ideal rate of addition of urea that produces the shortest reaction time with no danger of overheating.

The reaction is programmed in about 32 consecutive stages, which are set up as look-up tables in the memory of the control computer. There is a target rate of addition and a target temperature for each stage. Urea is added at a high rate to begin with, but the rate is tapered off as the reaction proceeds. At each stage, the kettle is weighed (by load cells) and the weight of its contents compared with the look-up table.

Its temperature is measured at the same time.The rate of feed of urea and the rate of flow of cooling water are adjusted accordingly. In the previous example, the rate of acceleration of the spoiler could be calculated beforehand, using the simple Newtonian equations of motion. In this chemical example, there is no simple equation. The targets are determined from experience gained when the plant was operated manually.

A complex control system

An example of a complex system with several interacting control loops is found in the steam generation plant at an electrical power station. The description below is based on the power station at Ironbridge in Shropshire, which is a coal-fired station.

The control of power generation at the station depends on the control of the boilers driving the turbines. The system operates to ensure that the steam delivered to the turbines is at a temperature of exactly 566°C and at a pressure of 150 bar. This is achieved by a computer-based system known as ABC (Automatic Boiler Control). The computer is programmed with algorithms to provide full P+I+D control. Fig. 2.12 represents the main elements of the system.

ABC comprises five control systems, all of which interact to varying degrees. These are:

* **Feed control** The flow of feed water to the boiler is controlled by varying the speed of the boiler feed pumps and the apertures of the six feed regulator valves.

* **Superheater temperature control** Controls 4 cooling spray valves in the desuperheater.

* **Fire control** Controls fire rate of the mills (where the coal is ground to dust), through six mill output controls.

* **Load control** Controlling the electrical load placed on the generator.

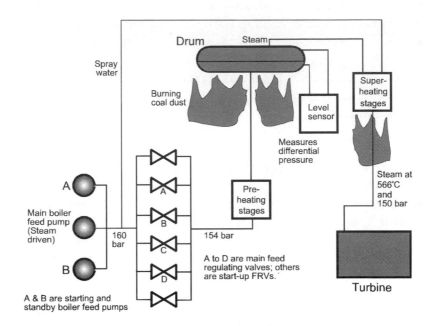

Figure 2.12 *The steam generation plant at Ironbridge Power Station comprises several inter-related control systems.*

- **Fan controls** There are two control systems under this heading. The *induced draught system* draws air through the furnace by expelling it after combustion. The *forced draught system* impels air into furnace before combustion.

We will look at some of these systems in more detail.

Feed control Water is pumped into the system by three pumps. The main boiler feed pump is steam driven, and is used when the generator is running at 200 MW or more. The A and B pumps are used for start-up (when there is no steam to drive the main pump) and for standby. These are electrically driven and have half the capacity of the main boiler feed pump. The output from these pumps goes to a common line in which the water pressure is 160 bar. The water is distributed to six feed regulating valves. Two of these are start-up

Differential pressure sensor

The level of water in the drum (Fig. 2.12) is estimated by measuring the difference of pressure at two levels within the drum. The difference of pressure is proportional to the depth of water. This is because the pressure at the upper level is practically equal to the pressure at the water surface and the pressure at the lower level depends on the depth of water and the known density of water at that temperature. There is a possible source of error because varying amounts of water condensing in the upper pipe introduce pressure differences along the upper pipe. This leads to errors. The solution is to keep the upper pipe full of water, so there is a *constant* pressure offset to be allowed for when measuring the pressure difference.

FRVs the other four (A to D in Fig. 2.12) main FRVs. The flow of water to the boiler is controlled by varying the speed of the pumps and by adjusting the valves (and the number of valves that are open). After the valves, the water enters another common line and is at a pressure of 154 bar.

The water passes through a number of pre-heater stages to the drum, where it is further heated and becomes steam. The action of the feed control loop is to maintain the water level in the drum. The sensor measures the pressure differential between two levels in the drum, one above the usual water level and one below it (see box).

Superheater temperature control After leaving the drum, the steam is passed to a superheater to raise its temperature to 566°C. In practice, the technique is a little more complicated than this. First, the steam is heated *above* 566°C, and then it is cooled to the correct temperature by spraying water into it in the desuperheater. The temperature of the steam coming from the desuperheater (Fig. 2.13) is

Control systems at work – 2

The originally hardwired control room at Ironbridge Power Station in Shropshire is now replaced by a fully computerised installation. Each control console has a bank of four monitors placed side by side. There are no control buttons or meters, and no computer keyboards All the hardware of a conventional control room has been replaced by the displays on the monitors. These show the state of every part of the steam generation plant, the turbines, the boiler and the electric power generator. The buttons and knobs that control the system appear as images on the monitor screens.

The operator has nothing more than a mouse with which to control the whole power station. Using techniques with which users of Windows are familiar, the operator uses the mouse to click on the images of buttons and so controls the power station. The reason for having four monitors is that there is not enough room on a single screen for all the controls that the operator may need to keep in view at one time. The use of the mouse extends to all four screens. As the mouse pointer moves off-screen it reappears on the next screen along the line.

Even with this high technology, there is still a conventional control panel situated in a corner of the control room. This has conventional push buttons; some of them enclosed in small protective boxes so that they can only be used with deliberation. The meters are of the traditional electromagnetic moving-coil type. Should anything go wrong with the computer control system, the major functions of the power station can instantly come under manual hardwired control.

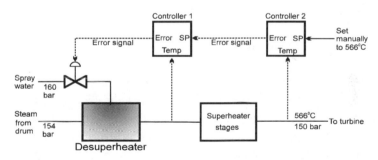

Figure 2.13 *Control of steam temperature at Ironbridge Power Station depends on cascaded control loops.*

monitored by a control circuit (Controller 1) which operates the water valve for spray water. This water is taken from the feed system between the feed pumps and the regulating valves, so its pressure is 160 bar. This means that it can be sprayed into the desuperheater, where the pressure is lower. The set point (SP of Controller 1 in Fig. 2.13) for this control is variable. After the steam leaves the desuperheater, it passes through further heating stages (plateau superheater and secondary superheater), emerging at 566°C and 150 bar.

The temperature as it emerges and goes to the turbine is monitored by another sensor and control circuit (Controller 2). The set point of this circuit can be adjusted by an operator. The error signal from this circuit is fed back to adjust the set point of the first temperature controller. Thus, we have two feedback loops *in cascade*. The main loop time delay (desuperheater to furnace) is just over a minute. The use of two loops in cascade, with one loop controlling the set point of the other, gives much more precise control of the steam temperature than could be obtained by using a single loop to heat the steam directly to 566°C.

Fan controls The forced draught fans produce a current of air to maintain a ball of flame where the coal dust is being burnt in the main part of the furnace (that is, around the drum and superheater). Hot gases flow on past the drum to heat the water as it flows toward the drum in the *economiser*. This is one of the pre-heating stages (Fig. 2.12) and is a way of making use of the heat that has not been used for heating the water in the drum.

The oxygen content of the gases is measured at the exit from the furnace. A control system operates to vary the rate of airflow to maintain the oxygen content of the exhaust air at a minimum of 3%. This ensures that the oxygen content at the combustion stage is enough to promote efficient combustion.

Fire control Coal is pulverised in the mills. These are drums in which the coal is mixed with 2-in diameter steel balls. The drums rotate, grinding the coal to dust. The dust is then blown into the furnace by the forced draught fans.

The ABC is an example of multivariable control. The system is very complicated, as there are so many interactions between the control systems of the various stages of steam generation. Additionally, the relationships vary according to the amount of electric power being generated at the time. Computer models of the individual systems have been produced to allow operators to study how the systems interact and how any changes in one loop affect other loops. Attempts can be made to balance the whole system by altering set points in the various loops. However, there are some unknown effects at work. For instance, it is difficult to measure the energy input to the system provided by a particular mix of air and coal dust. Such factors lead to inaccuracies in the model.

Test yourself

1 The difference between the output of a system and its set point is called:

    A hysteresis.
    B offset.
    C feedback.
    D error signal.

2 In a proportional control system:

    A the set point equals the output.
    B there is an offset between set point and output.
    C there is small open loop gain.
    D reducing the gain reduces the offset.

3 To minimise the offset error in a proportional system, we:

    A make the open loop gain as large as possible.
    B make the open loop gain as small as possible.
    C introduce negative feedback.
    D change the set point manually.

4 A bang-bang control system:

    A is rarely used in industrial plant.
    B has a large offset error.
    C has no feedback.
    D turns the actuator fully on or fully off.

5 The size of the actuator signal in a proportional + integral system does not depend on:

  A the difference between the output level and the set point.
  B the open loop gain.
  C the rate of change of input.
  D the time integral of the offset.

6 A P + I + D system:

  A is unaffected by noise in the system.
  B has zero offset error.
  C in inherently very stable.
  D shows second-order lag.

7 For the most precise response to changing input, we use:

  A a proportional system.
  B an on-off system.
  C a three-term system.
  D a P + I system.

8 First-order lag is usually characteristic of systems that have:

  A feedback.
  B inertia.
  C electronic noise.
  D offset error.

9 If a system has second-order lag, a large change in the input may cause:

  A oscillations in the output.
  B transit delay.
  C underdamping.
  D large offset error.

10 The effect of large input disturbances may be reduced by:

  A overdamping.
  B derivative control.
  C feedforward.
  D cascaded loops.

3
Sensors and actuators

This chapter describes a selection of the sensors and actuators most commonly used in control systems.

Sensors are sometimes referred to as *transducers,* a term which has an entirely different meaning. A transducer is a device that converts one form of energy into another form. An electric bell is a transducer because it converts electrical energy into sound energy (which is a form of kinetic energy). Similarly, a Ni-Cd cell is a transducer, converting chemical energy into electrical energy. However, neither of these are sensors. Conversely, a light-dependent resistor is not a transducer because the property that varies with light is its resistance, which is not a form of energy. Of course, there are devices that are both sensors and transducers at the same time. A piezo-electric microphone is an example. It can convert mechanical vibrations into a varying current and, in other circumstances, it can convert a varying current into mechanical vibrations. In the first instance, it is a sensor, in the second instance, it is an actuator, and in both instances, it is a transducer.

Some, but by no means all, of the sensors described in this chapter are transducers. All electrically powered actuators are transducers, so the use of the term *transducer* fails to distinguish between sensors and actuators, a distinction that is fundamental in control systems.

Sensors

The first group of sensors have an output that is binary. That is to say, if the sensor is or includes a switch, the switch is either open or closed. If the output is a voltage, it is either high or low. In some cases, the sensor itself has a voltage output that can take any value within a given range. In other words, it has an analogue voltage. This is converted to binary by an interface circuit that gives a binary high/low output depending on whether the analogue voltage lies above or below a given threshold value.

Switches

Perhaps the most frequently used sensor in industrial plant is the *limit switch*. It is often a microswitch actuated by a lever. Its function is to detect whether or not a part of the machinery has moved beyond a given limit. This may often be connected with making the machine safe to use, or to prevent it from damaging itself. The other main application is to detect if a workpiece is in position ready for machining. The limit switch may be normally open but is closed when the workpiece or a part of the machinery moves beyond a given limit, touches against the lever and trips the switch to close. The state of the switch, open or closed, is simple to detect electronically and to convert to a binary 0 or 1. A binary input such as this is easily processed by a logic circuit or by a computer.

The function of a limit switch is, as its name implies, to indicate when an object has moved as far as a given limit. A related type of switch is the *proximity switch*. As with the limit switch, it detects when an object has reached a certain position. The difference is that the object does not actually have to touch the switch in order to register its position. A proximity switch operates at a distance, though often this distance is only a few millimetres. For metal objects, we may use an inductive proximity switch, usually with a permanent magnet close to it to generate a magnetic field. The switch consists essentially of a coil wired into a balanced bridge circuit. As a metal object comes close to the switch, the resulting changes in the local magnetic field alter the strength of the field threading the coil, unbalancing the bridge, and generating a pulse. Capacitive proximity switches may also be used, particularly for detecting non-magnetic objects, as in Fig. 3.1.

Optical sensors

Optical sensors are often used for the same purpose as limit switches and proximity switches. In these applications, the sensor (usually a reverse-biased photodiode) is illuminated by a beam of light, often from an infrared LED or a laser diode. Normally, there is a leakage current through the photodiode but, when the beam is broken, the leakage is much reduced. Again, the change is easy to detect, and to interface to a control circuit. An example of the use of this type of sensor appears in Fig. 3.1.

Optical sensors can also be used in counting. As manufactured objects pass along the production line, they break the light beam. The pulses so generated are counted by a logic circuit and the total number of objects is recorded.

If manufactured objects are being stacked in layers by a machine prior to packing, the height of the pile may be monitored by mounting an optical sensor level with the intended top of the pile. When the correct number of objects has been stacked, the beam is broken and the pile is passed on to the packing stage.

Thermal switches

Thermal switches change their resistance markedly for a relatively small change of temperature. They have a fixed temperature at which this change occurs. For example, the resistance of one type of thermal switch changes from 100 Ω to 100 kΩ as its temperature passes through 75°C. A change of this size is easy to use.

Uses for sensors

Fig. 3.1 illustrates a machine used in making cardboard boxes. Its function is to apply a strip of glue parallel to one edge of a piece of cardboard. The card is fed in at one end of a narrow track and is glued as it passes through. The stages of the operation are:

1 Air is blown on its top and bottom surfaces to remove dust.

2 A fibre-optic sensor, which senses when there is card in the machine. Fibre-optics is used because there is not enough space in the machine to allow a straight optical path from lamp to sensor. The ends of the sensor also have air blown across them to keep them dust-free.

3 The card passes under a wheel which holds it firmly against the bottom of the track while the glue gun comes up from below (in a cradle moved by hydraulic pistons). It deposits a line of glue as the card passes through. The information from sensor 2 indicates when the glue should be applied.

4 The orifice of the glue gun is fine and must not be allowed to clog so, as the strip of glue is completed, the gun is retracted and a wiper moves across the end of the orifice. It cleans it and stays there to seal it until the next application.

5 Another blower cleans the card.

6 The card passes beneath a capacitative proximity detector. This is not a simple one-bit sensor but detects three conditions: no card, card only, card plus glue. This allows fault conditions to be detected.

7 At the rear there is an inductive proximity detector to detect the metallic glue gun. The gun can be fired only when it has been detected, that is, when it is positioned in its cradle, ready to squirt glue on to the card. This prevents the gun from being fired at other times, for example, when it has been removed from the cradle (which is potentially dangerous).

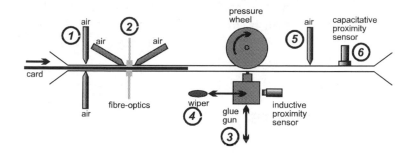

Figure 3.1 *Many complex industrial processes may be broken down into a sequence of relatively simple operations, each monitored by a simple sensor, and co-ordinated by a computer program.*

Instead of a fixed-point thermal switch, we may prefer to use a thermistor circuit, as in Fig. 1.11. Like the switch, the circuit has binary output. However, with the analogue sensor, we are able to adjust the circuit so that the circuit is triggered at any set point within the range of the thermistor.

Keeping up?

1 Define (a) a sensor, and (b) an actuator.

2 Describe two examples of sensors with a binary (on-off) output.

3 What type of sensor would you use for counting objects passing by on a conveyor belt?

4 What type of sensor would you use for controlling the temperature of a greenhouse?

The second group consists of sensors with an analogue output that is not converted to a binary form. Such sensors are essential in a system that has proportional control. Then we do not merely want to know if a given value is above or below the set point but by how much it differs from it. The range of analogue sensors is so vast that here we can describe only a small fraction of them. They can be divided into groups depending on the quantity to which they are sensitive. The more important groups are:

Position sensors

In many fields, from automatically operating a woodturning lathe to aiming a radio telescope at a particular part of the sky, it is necessary for a control system to know where things are. For this reason, position sensors (sometimes called displacement sensors) are vital to many control systems.

There are two types of position that may be measured and controlled:

Linear position: the location of an object (such as a cutting tool) along a straight path, measured from a point on the path.

Angular position: the direction of a lever-like object (such as an arm of a robot), measured as an angle from some reference direction.

One of the simplest forms of linear position sensor is essentially a variable resistor (Fig. 3.2). The object whose position is to be

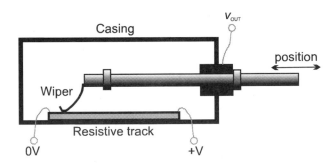

Figure 3.2 *In a linear position sensor the wiper is moved along a conductive track and v_{OUT} varies with the position of the plunger.*

monitored is coupled to the end of the plunger. Either there is a mechanical link, or the plunger is loaded by a spring and the object is pressed against the end of the plunger so that both move as one. As the plunger moves in or out of the casing, the wiper moves along the resistive material. There is a voltage drop along the resistive material and v_{OUT} varies from 0 V to +V as the wiper moves. The output voltage is thus directly related to position. The voltage may be used in a simple driver circuit or fed to an analogue-to-digital converter in a computer-controlled system.

A position sensor of the type shown in Fig. 3.2 has the advantage that its output is easy to read and to convert to binary form. One disadvantage is that the sensor prevents the object from moving freely because of friction in the bearings and between the wiper and the resistive strip. Another disadvantage is that eventually the strip wears thin in places, introducing errors into the system.

Another commonly used position sensor is the *linear variable differential transformer* or LVDT (Fig. 3.3). This is relatively expensive and the motion of the plunger is limited to a few millimetres. However, it does not suffer the disadvantages of friction or wear. Its main disadvantage is that a more complex circuit is needed to analyse its output. An alternating current (v_{IN}) is applied to the central excitation coil and this induces a magnetic field in the metallic core. The core couples the excitation coil to the two sense coils A and B, and currents are induced in them. The core is able to slide freely so that one coil is more strongly coupled to the excitation coil. A larger current is induced in that coil. The coils are connected so that, from the amplitude and direction of the induced current (v_{OUT}) the position of the core, and hence of the object attached to it, may be determined.

Although LVDTs have traditionally been the most frequently used position sensors, a more recent development is the *linear inductive position sensor* (Fig. 3.4). A centre-tapped coil provides two branches of a bridge circuit, the other two branches being provided by a pair of capacitors. A 1 MHz signal v_{IN} is fed into the bridge and keeps it

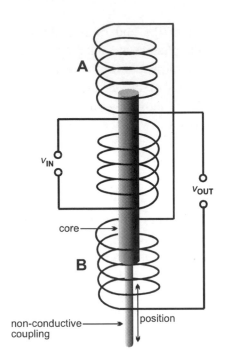

Figure 3.3 *In this drawing of an LVDT, the excitation coil is coupled more strongly with coil B than with coil A.*

resonating. An AC signal v_{OUT} is taken from the centre tap. The amplitude of this voltage depends on the relative inductance of the two halves of the coil. This depends on the position of the target, which can slide along inside the coil. The target is conductive (often aluminium) but non-magnetic and the alternating magnetic field within the coil induces eddy currents in it. These affect the inductance of each half of the coil to different degrees, depending on what fraction of the target is within each half. The output v_{OUT} is sampled at the same phase in each cycle to provide a DC voltage signal the magnitude of which is related to the position of the target.

An advantage of the LIP is that the sampling interface is relatively simple and can be fabricated as an integrated circuit. This gives it small size so that the interface can be incorporated within the sensor. With a 5 V supply, the output ranges from 0 V to 5 V as the target moves over its full range of travel. The coil of an LIP has only about a hundred turns, compared with several thousand for LVDT coils, so the sensor is cheaper and more robust. In addition, the LIP is able to measure displacements of 1 mm to 1500 mm. This compares with a stroke of just over 10 mm for a resistive position sensor. With an LVDT, typical displacements are only a few millimetres, the long-range types having a maximum of about 50 mm and a coil length of over 5 times that amount.

Optical linear position sensors are of two main types. One type uses the Gray code (see box) to give an absolute measure of position. The other type measures relative displacement. A position sensor based on a 4-bit Gray code is shown in Fig. 3.5. This is a 4-bit system, allowing 16 different codes and therefore locating the position of the encoder to 16 positions along its line of travel. The encoder consists of a transparent glass or plastic plate on which is printed an opaque pattern

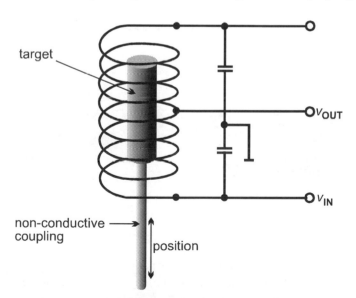

Figure 3.4 *The amplitude of the output from a linear inductive position sensor depends on the position of the target within the coil.*

Gray code

This is a sequence of binary code groups in which only one digit changes as we move from one code to the next in the sequence.

Example: 0000
              0001
              0011
              0010
              0110
              0111
              0101
              0100

This sequence continues for 16 groups, then repeats.

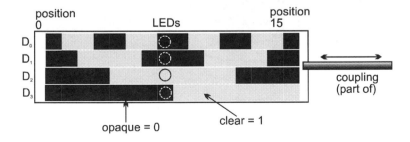

Figure 3.5 *When the encoder is at position 7, as in the figure, the code read by the optical system is 0100. If the encoder moves to position 8, only one digit, D_3, changes and the code becomes 1100.*

to represent the Gray code. A vertical row of four LEDs projects four beams through the plate, and there are four photodiodes on the other side to detect whether the area of the plate is transparent (=1) or opaque (=0). The point about the Gray code is that only one digit changes at a time. This removes any possible ambiguity as the plate moves from one position to the next. Until the changing digit has changed, the old code is read. After the change the new code is read. There are no intermediate states. Most encoders have more than four bits to allow the position to be read more accurately. Commonly available encoders have 12 bits, which allows for 2^{12} (= 4096) positions.

The second type of optical position sensor has a transparent plate marked with equally spaced bars (Fig. 3.6). As the plate moves, each bar in turn breaks the light beam, and the resulting pulses are counted. The distance moved is proportional to the number of pulses recorded. This tells the control circuit how far the plate has moved, but does not tell it in which direction. Directional information is obtained by

Figure 3.6 *A bar encoder has a plate with alternate opaque and transparent bars.*

placing a diaphragm in the optical path to obtain a signal that is 90° out of phase (Fig. 3.7). As the bar moves from right to left, the change from 00 to 11 passes through 10. As the encoder moves from left to right the change from 00 to 11 passes through 01. Simple logic can tell the control circuit which way the encoder is moving.

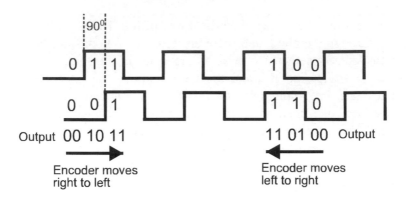

Figure 3.7 *The output of a bar encoder consists of two pulsed signals, one shifted 90° with respect to the other.*

Angular position

The descriptions of position sensors all refer to *linear* position sensors. Angular position sensors are equally important, particularly in controlling robot arms. All of the types of linear sensor described here are also available in rotary versions.

Speed sensors

There are a number of linear speed and acceleration sensors mostly depending for their action on the inertia of a seismic mass (a seismometer works on the same principle). However, when a system is computer-based it is generally simpler to measure *position* at intervals of time and calculate speed and acceleration from this data.

In control systems, we are more often interested in measuring angular or rotary speed rather than linear speed. A device for measuring angular speed is known as a *tachometer*. One of the simplest techniques is to mount a gearwheel on the shaft and use a Hall effect

Steering a telescope

The Lovell radio telescope at Jodrell Bank is steered to its required direction with the help of optical encoders. These are rotary versions of the encoders described in this chapter. There are two Gray encoders for measuring elevation, one situated on each tower. The mean of the two readings from these is taken as the elevation of the telescope.

There are three encoders for azimuth (angle with respect to North). One of these is a Gray encoder and measures the absolute azimuth. The other two are bar encoders, which give more precise azimuth measurements. As the telescope is moved to change its azimuth, a gear system spins a disc marked with 500 bars. Optical devices count the bars as they pass through the beam.

Signals are produced in quadrature, as in Fig. 3.7 to allow the direction of motion to be detected. Logic within the control system produces a 21-bit code covering the whole 360° of azimuth. Thus, 1 bit represents $360/2^{21}$ degrees of arc, which is equal to 0.6 seconds of arc. This is the angle subtended by a 10p coin at a distance of 8 km.

(magnetic) device or an inductive proximity sensor to produce a pulsed signal as the teeth pass by the sensing head. At low speeds, the individual pulses may be counted by logic circuits. At higher speeds, we generally use a special IC that converts frequency to voltage. Another type of tachometer is simply a DC generator, the output voltage of which is proportional to angular speed.

Level sensors

These are often position sensors adapted to measuring the position of the surface of a liquid. In their simplest form, we use a float connected through one or more levers to a linear or angular position sensor. Since the pressure in a liquid is proportional to depth, a pressure sensor (see next section) may be used for measuring level. With certain liquids, a capacitative position sensor placed above the surface of the liquid is used to measure the distance between the sensor and the surface.

Mechanical force

Force is measured in two principal ways, by strain gauge or by piezo-electric means. A strain gauge consists of a thin metal foil forming a number of narrow parallel conductors, joined at their ends to make a continuous zigzag conductor. The foil is mounted on a self-adhesive plastic backing. This is fixed on to the surface of a beam, girder or other object to measure the distortion of that object when it is subjected to mechanical stress. An example of such an object is the *load cell*. The kind illustrated in Fig. 3.8a has four strain gauges attached to it. When a gauge is under tension the conductors become longer and narrower and their resistance increases. Its resistance decreases when it is compressed. The four load cells are connected into a bridge, which is balanced when there is no strain in the load cell and the resistances are equal. The voltage output from the bridge is zero. When forces act on the load cell to change its shape (Fig. 3.8b), the resistance of two of the gauges is increased and the resistance of the other two is decreased. This throws the bridge out of balance and an output voltage is generated, proportional to the force.

Load cells are often used for weighing heavy objects, such as loaded trucks on weighbridges. However, small load cells are made for weighing objects less than 2 kg. Weight is an important quantity,

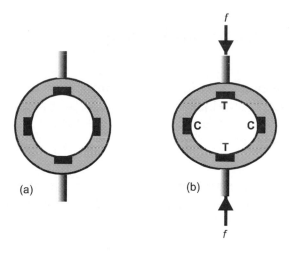

Figure 3.8 *One form of load cell (a) is a metal ring with four strain gauges mounted on it. When external forces (such as weight) compress the ring, two of the gauges are under compression (C) and two are under tension (T)*

especially in the dispensing of fluids and powders and load cells are used in controlling automatic weight (see box overleaf).

A strain gauge may also be used in pressure sensors. A typical pressure sensor has two chambers separated by a diaphragm. One chamber is connected to a part of the equipment in which the pressure is to be measured. The other chamber is connected to a vacuum (for absolute pressure measurement) or to a fluid under a reference pressure (for relative pressure measurement). The difference between the pressures on the two sides of the diaphragm cause it to bulge one way or the other. There are one or more strain gauges mounted on the diaphragm to detect the amount by which it bulges, and the direction. The voltage produced by the sensor is proportional to the pressure difference. Pressure sensors are made that operate over various ranges, covering a few millibar up to several hundred bar.

Load cells for weighing

Light-duty load cells are used in a production line for filling plastic cartridges with toner for copying machines. As the cartridges are moved along the conveyor belt they pass through these stages:

1 A load cell is used to measure the tare (empty) weight of the cartridge. The weight is recorded in the memory of the controller.

2 An auger forces toner into the cartridge until it is about 80% full. The quantity of powder delivered is determined by the speed of rotation of the auger and the time for which it operates.

3 The cartridge is shaken to make the toner settle.

4 A second auger forces more toner into the cartridge to fill it.

5 A second load cell weighs the cartridge again. Its full weight is compared with the tare weight recorded further back along the line.

6 If the cartridge contains insufficient toner, a pusher arm ejects it on to a platform at the side of the line. It is taken manually, emptied, and returned to the beginning of the line.

7 The full cartridge goes on to be capped and sealed.

Note that, although it would be possible to design a control system to ensure that the cartridge always receives exactly the correct charge (within limits), this is too expensive to set up. It is more economic to employ a simple filling system and reject the faulty cartridges.

Control systems at work – 3

The packing line at a cheese factory makes use of a load cell to check the weight of the cheese slabs prior to wrapping. Slabs are carried from the cutter by a moving belt (1). They pass on to a short segment with its own moving belt. This is mounted on a load cell (2).

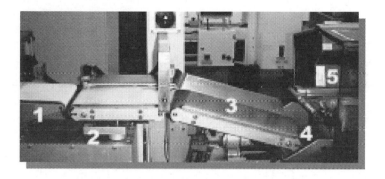

If the slab is not within the specified weight limits, the next segment of the line (3) hinges down, allowing the slab to fall down a chute (4) into a reject container. If the slab is within limits, the hinged segment rises flush with the next section of the line (5). This conveys the slab to the wrapping and labelling stages.

An ingenious use for pressure sensors is found in certain robot arms. The robot arm is used for handling plastic castings and placing them in a machine that glues several castings together. The arm has several suction pads on it, located to contact suitable points on the casting. A vacuum pump generates the suction and, when all pads are in contact with the casting, this holds the casting firmly at the end of the arm. To let go of the object the vacuum is switched off. A pressure sensor in the arm measures the air pressure at the suction pads. If the vacuum pump is on, but there is no vacuum at the pads, this means that there is no casting present, for air is able to rush freely into the suction pads. The process is halted until pressure at the pads drops, indicating that the arm has gripped a casting. In effect, the pressure sensor is acting as a proximity detector. This is but one example of the many unusual ways of using sensors in control systems.

Flow sensors

Measuring the flow of liquids and gases is an important aspect of control systems. We may need to measure flow of combustible gases to furnaces, the flow of coolants, or the flow of various reagents in chemical plant. The most commonly used technique is to use a rotating vane inside the pipe through which the fluid is passing. The rate of rotation is proportional to the rate of flow. Various methods are used for measuring the rate of rotation. In some flow sensors, a beam of light is passed through the pipe to a light sensor placed on the opposite side. The beam is broken as each blade passes through it, producing a pulsed signal from the light sensor. This is easily converted to a measurement of the flow rate. In other flow sensors, the vane is made of stainless steel and a magnetic sensor outside the pipe detects the blades as the vane spins. A pulsed signal is generated which can then be processed to measure the rate of flow.

Another commonly used method of measuring flow depends on the reduction in pressure that occurs when the flow passes though a narrower tube. This is the *venturi,* which is also used for measuring wind speed and the airspeed of aircraft (Fig. 3.9). A differential pressure gauge measures the reduction in pressure, which is proportional to the square of the flow rate.

An ingenious technique for measuring flow depends on Corioli's Force. This is the force that causes a vortex when water is draining from a bath or washbasin. The flowing liquid is passed through a metal tube shaped is an almost complete circle, like the Greek letter Ω.

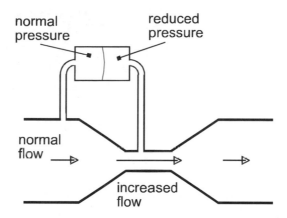

normal
pressure

reduced
pressure

normal
flow →

increased
flow

Figure 3.9 *A venturi tube produces a difference in pressure that is proportional to the square of the rate of flow.*

Corioli's Force distorts the tube by an amount proportional to the rate of flow. The distortion is measured by strain gauges mounted on the outside of the tube. Here there is the advantage that the liquid is totally enclosed in a metal tube. This technique is particularly suitable for flammable or highly toxic liquids, provided that they do not corrode the tube.

Temperature

So many processes depend on temperature that a wide range of temperature sensors has been produced, depending on a variety of physical principles. A *thermistor* is a sintered mixture of sulphides, selenides or oxides of nickel, manganese, cobalt, copper, iron, or uranium. It has the property that its electrical resistance varies with temperature. Most thermistors (ntc types) have a negative temperature coefficient and their resistance decreases with increasing temperature. Others have a positive temperature coefficient (ptc) but these are more often used for other purposes such as limiting current flow in a circuit.

In a sensing circuit (see Fig. 1.11), a thermistor is generally part of a potential divider. As temperature increases, resistance decreases, and the potential across the thermistor decreases. The potential is fed to a transistor, which may be connected as a switch, to turn on a heater, for example, or to activate a relay, at a preset temperature. It may also be fed to an op amp for similar processing. In circuits with analogue output, the potential is treated as being proportional to temperature. Unfortunately, the relationship is not quite linear, but this is usually not of importance over a limited range of temperatures.

Thermistors are made in various forms, including discs, bars and beads. Beads may be naked or sealed in a glass or plastic capsule to protect them from corrosive liquids. Thermistor beads can be very small, which gives them two advantages: they have low heat capacity and so quickly reach the ambient temperature, and they can be used to measure temperatures within very restricted volumes. Thermistors may be used to measure temperatures in the range from −50°C to +300°C.

Thermocouples are used for sensing high temperatures. Their operating range extends up to 1000°C or more. A thermocouple consists of a junction of two dissimilar metals or alloys. Usually it is made by twisting two wires together and then welding them. The K-type thermocouple, for example, consists of a junction between a nickel-chromium alloy and a nickel-aluminium alloy, and has a range from 200°C to 1100°C. For temperatures up to 1350°C, we use a type R thermocouple, which is a junction between a rhodium/platinum alloy and pure platinum.

A thermocouple consists of two junctions, the 'cold' junction (often at room temperature, or for greater precision at the temperature of melting ice) and the 'hot' or sensing junction (Fig. 3.10). When the junctions are at different temperatures an e.m.f. is generated which is proportional to their temperature difference. The e.m.f is of the order of a few millivolts so an operational amplifier circuit is nearly always essential for measuring its output.

By making the junction from fine wire, or small pieces of thin foil, the sensing thermocouple can be made very small and with low heat

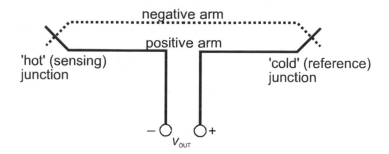

Figure 3.10 *In a Type K thermocouple, the positive arm is made from nickel-chromium alloy and the negative arm from nickel-aluminium alloy.*

capacity. By contrast, a heavy-duty thermocouple of rugged construction can be made from thick wire.

A *band-gap sensor* is a very convenient and precise device for measuring temperature electronically. In appearance, it looks like a transistor, with the usual three terminal wires. It comprises a silicon chip bearing two current-generator circuits. These circuits act together so that the voltage between the output pin and the ground pin is proportional to the temperature. The balance is such that a change of temperature of one Celsius degree causes a change in voltage of exactly 10 mV. This equivalence makes a band-gap sensor easy to calibrate. The range of the device is 0°C to 110°C with an accuracy of 0.4°C in the centre of the range.

Keeping up?

4 What is the advantage of using an analogue sensor instead of one with on/off action?

5 Explain how a rotary variable resistor may be used as an angular position sensor.

6 Describe the action of a sensor based on induction.

7 Explain the nature and purpose of the Gray code.

8 What is a strain gauge? Describe how it is used for measuring (a) weight, and (b) pressure.

9 Name some temperature sensors. Which type would you prefer for measuring the temperature of an oil-bath?

Actuators

Apart from heaters and lamps, the most often used actuators are those that provide some kind of force or (its consequence) motion. We use actuators to punch holes, fix rivets, bend sheet metal, compress and melt plastic granules, cut strip plastic into lengths, move robot arms, turn lathes, position drills and cutting tools, rotate telescopes, eject faulty products, and fix labels to bottles. These are only a few of the tasks of actuators, and some more will be mentioned in this chapter.

The most basic type of actuator is the solenoid (Fig. 3.11), consisting of a coil and an armature. The coil has several hundred turns of fine wire and is wound on a former. The iron armature, usually chromium-plated, slides freely in and out of the coil. For normal operation, the armature is at rest with most of it projecting from the coil. When a voltage is applied to the coil, the magnetic field causes the armature to be pulled forcibly into the coil. Different types operate on a DC voltage between 6 V and 24 V, producing a pulling force ranging from

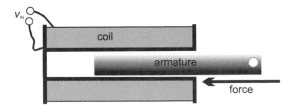

Figure 3.11 *A solenoid is used to produce a strong pulling force.*

a few hundred grams-force up to 4 kgf. This is ample for many machining and handling operations and the circuitry to drive a solenoid from a logic output is very simple.

The typical solenoid generates a *pulling* force and may rely on an attached spring to return it to position when it is switched off. A solenoid may also generate a pushing force by having a narrow thrust rod mounted on the end of the armature and projecting out through the other end of the former (that is, on the left in Fig. 3.11). Another variation is the latching solenoid, which incorporates a permanent magnet. When the coil is energised the armature is drawn into the coil and then comes into contact with the magnet. This holds the armature firmly in position when the current is switched off. This arrangement saves power and reduces overheating when the solenoid has to be energised for a long period. The latching solenoid is unlatched by passing a pulse of reverse current through the coil. Solenoids may also be operated on AC, such as the mains supply, if this is more convenient to provide than a low-voltage DC supply.

From simple solenoids, the next step is to couple the solenoid to a valve. A large range of solenoid controlled valves is made for controlling the flow of air, various gases and many different liquids. In their turn, air valves may be used to distribute compressed air to pneumatic actuators. These usually consist of small cylinders and pistons. Some have only one-way action so that opening the valve supplies air to the cylinder and the piston moves. A spring or gravity returns it to its inactive position when the valve is closed. Other cylinders have connections to both sides of the piston, providing for two-way action. A system that operates in this way is known as an *electro-pneumatic system.*

Electro-pneumatic actuators have the advantages of developing greater force and having a longer stroke than solenoids. They are often used in plant that are processing flammable liquids. With an electrical actuator, there is always the risk of sparks or of coils burning out or overheating, leading to a serious fire. The electro-pneumatic valves are situated in the control room some distance from the plant. Air lines connect the valves to pneumatic actuators on the plant itself.

Electro-hydraulic systems operate on similar principles. They are sometimes used in flight control of aircraft, for example. An electric pump is switched on to compress the hydraulic fluid which then moves pistons connected to the flight surfaces, such as ailerons and spoilers.

Control systems at work – 4

When on the ground, the Airbus A330 is steered by electrohydraulic actuators.

A hydraulic cylinder (1) is connected by a crank mechanism (2) to the vertically pivoted undercarriage assembly (3). The angular position of the assembly is sensed by a limit switch (4). This provides feedback to the steering control system.

A version of the solenoid known as a *rotary solenoid* is used to produce a limited rotary action, as required in certain operations. For unlimited rotary action, a wide range of motors, both DC and AC have applications in control systems. Often the motor drives a reduction gear system. A steering system, for example, may then be turned relatively slowly and over a limited angular range, but with high power.

Although a motor may be most simply controlled by switching its power supply on or off, and its speed may be controlled by varying the supply voltage, there are many instances when closer control is essential. For example, a motor may need to be run at high speed on some occasions yet to run smoothly at very low speed on other occasions. For example, at Jodrell Bank, control of the motion of the telescope has to cover a wide range of speeds. It needs to be turned at high speed for moving the telescope to point at different parts of the sky, and low speed, sometimes zero, when tracking an astronomical object.

Fig. 3.12 shows a very simple but effective circuit for controlling the speed of a small DC motor. The op amp monitors the voltage across the motor and acts to keep it constant, equal to the control voltage, v_c. This ensures constant current through the coils, keeping it turning at constant speed, in spite of variations in the load on the motor.

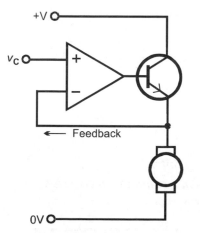

Figure 3.12 *An op amp circuit such as this one maintains a constant voltage across the motor. This compensates for fluctuating back e.m.f. as the load on the motor changes and so helps to keep the motor running at constant speed.*

Instead of controlling the amount of current supplied to the motor, the current can be supplied in pulses of constant amplitude but with varying mark-space ratio. The speed of the motor depends on the *average* current being delivered to it. In Fig. 3.13, the pulses are generated by a timer IC. The timer runs at 50 Hz or more, producing a stream of pulses. The inertia of the motor keeps it turning at a steady speed during the periods between pulses. Adjusting VR1 varies the pulse length and thus the mark-space ratio of the output to Q1. This technique gives smooth control of motor speed over a wide range. There is less tendency to stall at low speeds because the pulses switch the transistor *fully* on and the motor is driven with full power. It is possible for the pulses to be generated by a microprocessor for fully automatic speed control. The main disadvantage of the method illustrated in Fig. 3.13 is that it is an open loop. The actual speed attained depends on the load. However, there is no reason why feedback can not be added to provide full speed control.

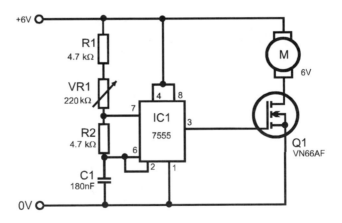

Figure 3.13 *The timer produces pulses of variable mark-space ratio. The longer the 'on' time, the faster the speed. Since all pulses are delivered at full voltage, the motor is always strongly driven, even when running slowly. This avoids the motor stalling at slow speed.*

Control of this type of system may use the special strategy illustrated in Fig. 3.14. This is known as a *servo system*. It is an on-off control system, with the extra feature that the motor is not simply switched on and off but may be made to rotate in either direction. The control circuit is built from op amps and transistors, or it may be a special IC designed for servo driving. If the object is not at its intended position, pulses are supplied to the motor with polarity such that it rotates in the appropriate direction until $v_C = v_S$.

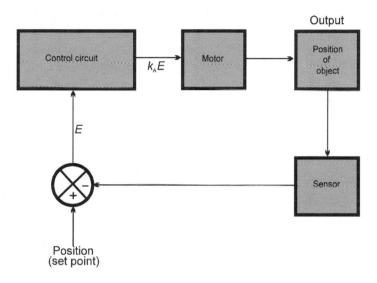

Figure 3.14 *In a servomechanism, a sensor monitors the position of the moving object. The error signal is proportional to the displacement of the object from its intended position and the polarity of the supply to the motor acts to reduce the error to zero.*

One of the possible snags of such a system is that the system overshoots. Instead of homing steadily on the intended position, the object is moved *past* that position. It is then moved toward the position from the opposite side, and may overshoot again. This may continue indefinitely, causing the object to oscillate about the position instead of settling there. This is called *hunting*. To avoid hunting, most servo systems have a dead zone or *dead band* on both sides of the intended position. If, for example, v_C is less than v_S and increasing, the motor is switched off as soon as $v_C - v_S$ becomes less than a given small amount. The object stops a little way short of the intended position but the error is small enough to ignore.

A *servomotor* is specially designed to provide feedback of angular position. It has reduction gearing so that the lever (Fig. 3.15) turns relatively slowly (typically taking 0.2 s to turn 60°) when the motor is running. If required, the angular motion of the lever can be converted to linear or other forms of motion by connecting it to suitable gearing or lever systems. The servomotor has a rotary potentiometer on the same spindle as the lever. If the ends of these are connected to 0 V and +V the output v_S from the wiper of the potentiometer is proportional to

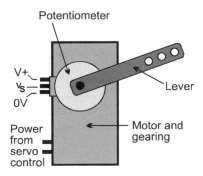

Figure 3.15 *A servomotor has a built-in potentiometer to feed back information about the lever position to the control circuit.*

the angular position of the lever. As in Fig. 3.14, a control circuit is used to make the motor turn to bring the lever to the intended position. Servomotors are often used in robots and in model aeroplanes and vehicles.

Another way of exercising fine control over the running of a motor is the *switched reluctance motor*. This has a rotor with several outwardly directed poles, often six as shown in Fig. 3.16. Around this is the stator with inwardly directed poles, greater in number than those of the rotor, typically eight. Coils are wound around the poles of the stator, opposite coils being wired together. If the position of the rotor within the stator is as shown in Fig. 3.16, and the current to the upper and lower pair of coils is zero, then a current flowing through any of the other pairs of coils will attract the rotor to turn. In operation, the current through the pairs of coils is switched on in sequence and in varying amounts to produce a rotating magnetic field that keeps the rotor spinning. By suitable control of the currents, it is possible to spin

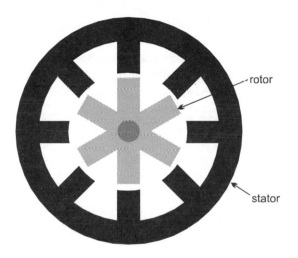

Figure 3.16 *A switched reluctance motor has more poles on its stator than on its rotor. Only one pair of poles on the rotor and stator can be aligned at any one time. The rotor, which does not have coils around its poles, is made to turn at a controlled speed by switching pulses into the coils of the stator.*

the rotor in either direction and at any speed. As might be expected, switching is a complicated matter. It is necessary not only to switch the current through each of the four pairs of coils on and off at exactly the right instants, but also to provide a pulse of exactly the right shape. If this is correctly done it is possible to spin the rotor very smoothly without any juddering.

A switched reluctance motor is controlled either by a special controller IC, or under the control of output from a microcontroller or microcomputer. The shape of pulse required and its timing varies with the speed of rotation of the motor. For this reason, the control system includes a tachometer attached to the shaft of the motor to measure the actual speed of rotation. It is also necessary to measure the angular position of the shaft as it rotates. The information is fed back in a closed loop to the control IC or computer.

With suitable control routines, it is possible to run a switched reluctance motor at any constant speed within its range, to change speed at a predetermined rate, and to make a smooth reversal of direction of rotation. In addition, the switching routine is designed so that the motor operates very efficiently and is economical of power. This is why this type of motor is so often used for driving actuators.

Keeping up?

10 Describe the action of a solenoid.

11 List two possible applications of solenoids in machine control.

12 What is a servo system?

13 Give an example of the way a servomotor is used.

14 Explain how a switched reluctance motor works, and why this type is so often used as an actuator.

15 Describe how a pulse generator can be used to control the speed of a motor.

Stepper motor

A *stepper motor* has a very similar construction to a switched reluctance motor. It has four sets of coils, arranged so that the rotor is turned from one position to the next as the coils are energised in a fixed sequence. The difference between a stepper motor and a switched reluctance motor is that the coils are switched fully on or off, making the motor rotate 15° at each step of the sequence. If the sequence is halted the motor halts instantly with its rotor 'locked' in position. Thus, the motor is useful for applications in which a shaft has to be turned to a number of fixed positions, situated in multiples of exactly 15° apart. The emphasis is on precision of position.

The control circuit may be built from standard logic gates or is available as a special IC, such as the SA1027. The rate at which it runs through the switching sequence depends on the system clock. If this is set to a definite frequency, the motor turns at a fixed rate, one revolution for every 24 pulses. Varying the clock rate varies the speed of the motor. There is no shaft-position sensor or speed sensor, so the motor runs in open-loop mode. The advantage of the stepper motor is that its speed is controllable with the same precision as the system clock. It is not affected by the load on the motor, except perhaps an excessive load, which might completely prevent the motor from turning. The stepper motor may be made to turn 7.5° per step by using a slightly different switching pattern. Motors with a 1.8° step angle are also produced. A related device is the *stepping linear actuator*, which moves a plunger rod a small distance at each step.

Test yourself

1 Which of these sensors has binary output?

        A A strain gauge.
        B A thermistor.
        C A thermal switch.
        D A light dependent resistor.

2 A device that detects whether a given kind of object is present or not is called a:

        A binary output device.
        B limit switch.
        C proximity switch.
        D tachometer.

3 An LVDT is used for measuring:

> A linear position.
> B magnetic fields.
> C the rate of rotation.
> D light values.

4 If the value currently shown on a Gray code encoder is 1101, the next value will be:

> A 1110.
> B 1101.
> C 1111.
> D 0010.

5 The number of positions of a Gray code encoder with 10 bits is:

> A 4096.
> B 1000.
> C 4095.
> D 1024.

6 A venturi tube is used to measure:

> A temperature.
> B air speed.
> C altitude.
> D air pressure.

7 The type of thermal sensor best suited to measure temperature in restricted volumes is a:

> A thermocouple.
> B band-gap sensor.
> C thermistor.
> D thermostat.

8 When alternating current is passed through the coil of a solenoid:

> A the armature is pulled strongly into the coil.
> B nothing happens.
> C the armature is ejected from the coil.
> D the armature vibrates.

9 With a switched reluctance motor we can control:

        A speed only.
        B speed and direction.
        C the back e.m.f.
        D speed, acceleration and direction.

10 A sensor made by twisting wires of two different metals together is called a:

        A thermocouple.
        B Hall effect sensor.
        C thermistor.
        D tachometer.

4
Linking sensors to actuators

A control system comprises one or more sensors, one or more actuators, and a control unit. The sensors provide the inputs to the system, the actuators give practical effect to the output. The control unit links the sensors to the actuators, exercising a control function as it does so. The control unit may be nothing more than a single transistor, which hardly deserves such an impressive name as 'control unit'. On the other hand, it may be a powerful computer that implements a complex control regime. In this chapter, we look at the more elementary control units, leaving microcontrollers and microcomputers until Chapter 5.

Elementary control systems are of three main kinds:

- Analogue controllers
- Relay logic
- Logic controllers

Early control systems were based on analogue circuits. Nevertheless, even in the early days before the development of logic ICs, there were digital circuits of a kind. These were based on relays, the on-off nature of relay operation providing a binary basis for the logic. Later, these were largely replaced by circuits based on logic integrated circuits.

Analogue controllers

An example of a simple analogue controller is seen in Fig. 3.12. The input is a voltage that can be varied in infinitely small steps over a given range. The output is the speed of the motor, which likewise varies smoothly over a given range. There is no stage at which the signals are digital.

The input to most systems is analogue, whether the system itself is mainly analogue or digital. This is because the world with which most systems interact is essentially an analogue world. Sensors detect temperature, position, or flow rate, all of which are analogue quantities. Actuators too are often analogue. For example, the motor control circuit we mentioned above has analogue input (control voltage), analogue processing (the op amp) and analogue output (motor speed). With switched reluctance motors it is possible to control angular acceleration as well. Servo systems are another example of analogue control. Apart from systems with an 'on-off' input (from a keyboard or limit switches, for example) and an 'on-off' output (relay energised or not, lamp on or not, valve open or closed) all systems are analogue at some stage.

Analogue systems are still in use in certain applications, such as central heating systems but, in more and more systems, the *processing* stage at least is being taken over by digital systems. It is in the *processing* stages that digital systems excel. A wholly analogue system may often with advantage be replaced by a system with:

- Analogue input stage.
- Analogue-to-digital converter (ADC).
- Digital processing/control stage.
- Digital to analogue converter (DAC).
- Analogue output stage.

Compact discs and digital video discs are part of such a system. The input stage (microphone or camera) is analogue and the output stage (speaker or VDU) is analogue, but the rest of the system is digital. This digital stage includes the manufacture and distribution of the discs.

The advantages of digital systems are so great that many systems that were originally analogue are being or have been converted to digital operation. For example, when the Lovell Telescope (then called the Mark 1 telescope) at Jodrell Bank began operations in 1957, its control was purely analogue. However, control has been largely digital since 1970. Input is digital (through keyboards), and processing is digital, but the later stages of output are analogue. Control of the position of the telescope is effected by two analogue drive voltages, one for each axis, each of which ranges from -10 V to +10 V. Given the required change of direction of the telescope, the drive rate is calculated by computer and a digital signal is fed to the control gear on the structure. There, DACs convert the digital signals to the analogue drive voltages. For servicing or in emergencies, it is possible to switch out the automatic drive voltages and use voltages produced locally by manual switching at stations on the structure.

The advantages of digital processing over analogue processing are as follows:

- **Precision:** The precision of voltages in analogue circuits depends on the precision of the components used to build the circuits. Typically, resistors are precise to 1%. High precision resistors with a tolerance of 0.01% can be obtained, but these are extremely expensive and made in only a small range of different values. Typical capacitors have a precision of only 5% and precision of 1% is regarded as high. By contrast a 12-bit digital system has a precision of 1 in 2^{12}, which is 1 in 4096 or 0.02%. For higher precision there is little problem in increasing the number of bits to 16 or more.

- **Accuracy**: The value of many components drifts in time, imparting a permanent offset to the voltages. Op amps are an essential component in many processing circuits, but these are subject to systematic errors such as input offset voltage. The accuracy of digital signals does not change.

- **Noise:** The noise inherent in electronic circuits and the risk of picking up electrical interference set limits to the amplification of signals, and to the reliability with which two signals can be compared. A digital circuit is not affected by noise of the types met with in analogue circuits. However, the step-like nature of digital signals may persist after conversion to analogue, causing *quantisation noise*. This can be ignored if the number of bits in the signal samples is sufficiently large.

- **Temperature:** The performance of many components (including resistors, capacitors and transistors) varies with temperature. Digital circuits are designed so as to be free from temperature errors.

A further advantage, which applies to the systems described in Chapter 5, but not generally to the simpler digital systems described in this chapter, is that digital systems may be:

- **Programmable:** When the parameters of a control system change, perhaps because of changes in manufacturing techniques, it is usually necessary to alter some of the component values in an analogue system. If the actual technique changes, it may be necessary to scrap the existing system and build a new one. This can be costly and may involve relatively long periods during which the system is out of commission. With a digital system minor or major changes in the system can often be accomplished simply by re-writing the software. Hardware costs are minimal or non-existent. The revised program can be written on a separate computer, then downloaded ready to run in an instant.

Keeping up?

1 Give an example of an analogue control system and a digital control system.

2 What comes between the input and the processing stages in a digital control system?

3 List five ways in which digital control systems are usually superior to analogue control systems.

Relay logic

This is logic performed by ordinary relays, occasionally supplemented by special types of relay and other similar electromagnetic devices. Although this technique is no longer favoured for new installations, there are still many older systems in use that operate in this manner.

Fig. 4.1 shows an example of a relay logic circuit, designed to carry out the logical function $Z = A$ AND (B OR C). The coils of the relays are shown unconnected in the figure but each would be connected to a DC power supply (say 12 V) through a separate switch. The switch could be of any type, such as a key, a push-button or a limit switch. In some applications it might be a circuit that produces a high output when a beam of light falls on a photocell.

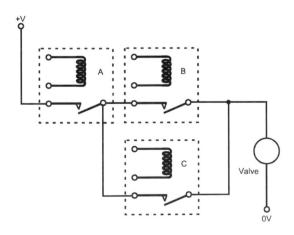

Figure 4.1 *Three relays wired as shown above perform the logical operation Z = A AND (B OR C).*

Examination of the wiring of the relay contacts reveals that current can flow from the positive supply (+V) to the solenoid actuated valve only when we energise relay A and either one (or both) of relays B and C. As a working example, consider that valve is on the outlet from a heating tank. The switch controlling relay A is a proximity switch detecting when a receptacle is present under the outlet to receive the processed liquid. Relay B is controlled by a thermal switch that closes when the liquid reaches a preset temperature. These two relays let the liquid warm to temperature, then open the valve to let the liquid drain into the receptacle, providing one is there.

Relay C is controlled by a manual switch used by the operator to drain the liquid at other times, again providing the receptacle is there. This simple system has obvious practical flaws. If the receptacle is not there and the operator does not turn open the valve manually, we need a relay to turn off the heater or to sound an alarm to prevent the liquid from overheating. As usual, a logical task needs a more complicated solution than is at first thought necessary. Unfortunately, if the snags of the system are thought of *after* the system has been designed and built,

Fig. 4.1 is based on relays with normally open contacts. Some types of relay have normally closed contacts, which break contact when the coil is energised. These allow the INVERT or NOT function to be implemented. Special relays are available with delayed contact closing to introduce an element of timing into a system.

Another device related to the relay is the uniselector. This is an electromagnetically controlled rotary switch. Each time a pulse is fed to its coil, a ratchet mechanism steps on, switching current to one of a number of output terminals. A uniselector can be used for controlling sequences of operation, and also for counting.

Further complexity in the logical operations of relays is obtained by wiring relays to switch other relays, or even to switch themselves. Self-switching is used in building latching circuits.

Relay logic in symbols

Relay logic is commonly set out as a diagram, using the symbols shown in Fig. 4.2. An example is Fig. 4.3, which is the symbolic equivalent of Fig. 4.1. This is much easier to interpret. It can be seen that current can flow from the left of the diagram to the valve solenoid on the right if a path is provided through relay A *and* through either of relays B *or* C.

(a)

(b)

(c)

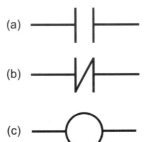

Figure 4.2 *These are the American-style symbols used to represent relay logic. They are (a) relay with normally open contacts, (b) relay with normally closed contacts, (c) an actuator.*

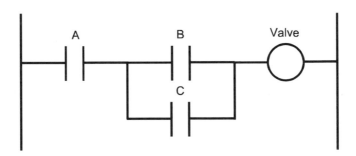

Figure 4.3 *When the circuit of Fig. 4.1 is re-drawn using relay logic symbols, it is much easier to understand how it works.*

One can buy a useful assortment of logic ics, or even a whole microprocessor, for the cost of one typical relay, so it is obvious that a relay logic system is more costly than its equivalent in other technologies. Moreover, a very wide range of complex functions is available very cheaply as integrated circuits. This is not true of relays, so building complex functions entails using more relays, taking up more board space and requiring more power.

Revising a relay system to modify its action is seldom simple. It is often more economical to scrap the board and build a new one.

Relays operate much more slowly than logic ics, though this is not always a disadvantage. A typical relay takes 10 ms to operate. A TTL logic gate takes only 10 ns, which is a million times quicker. The main advantages of relays are their reliability and that they can be used to switch heavy currents, including AC.

Switching of heavy current to inductive loads such as electric motors and solenoids may eventually result in contacts arcing and fusing together. In certain applications this may result in danger or damage. An example of such a fault is a motor that continues to run when it should be switched off. If safety is at risk, it is possible to use a relay that incorporates *forced operation*. Such relays have a second pair of contacts, which can not close if the first pair become welded together. Thus the forced operation relay switches of the machine as soon as there is a fault in the relay.

Keeping up?

5 Draw a logic circuit as in Fig. 4.3 to represent the logical statement $Z = (A$ AND $B)$ OR C.

6 State two advantages and two disadvantages of relay logic when compared with logic ics.

Logic controllers

By 'logic controller' we mean a system built mainly from logic ICs of the TTL, CMOS or, occasionally, the ECL and other families. Each of these families includes a wide range of logic integrated circuits. Some ICs provide simple logic gates, such as AND and OR. Others are able to perform complex operations such as counting. Wiring together these ready-made units is made easier by designing printed circuit boards to accept the ICs and provide the electrical connections between them. Once a design has been proved, the circuit can be relied upon to work reliably for a long time. Its logic can not be lost or corrupted. The circuit board of a logic controller is much smaller than that of the equivalent relay logic circuit.

Fig. 4.4 is the first draft of a logic system for controlling the entrance barrier of a car park. There are two phases of operation:

- Phase A: The barrier is down, a car arrives at the barrier, provided that there is a vacant space in the park, the barrier is raised.
- Phase B: The car drives under the raised barrier, the barrier is lowered.

After Phase B has ended the system returns to Phase A to await the arrival of the next car.

Phase A is controlled by the 4-input AND gate, labelled A in the diagram. A photo-sensor at the barrier detects the car and produces a high (1) output. If the barrier is down, a limit switch produces a high output. If there is a vacant space in the car park, the magnitude comparator detects that the number of cars in the park is less than the pre-set maximum number. It produces a high output. Finally the enable signal from the flip-flop is high when the system is in state A. With all its inputs high, the output of the AND gate A goes high. This causes pulse generator A to produce an output pulse. The pulse triggers a circuit that starts the motor to raise the gate. It raises the gate

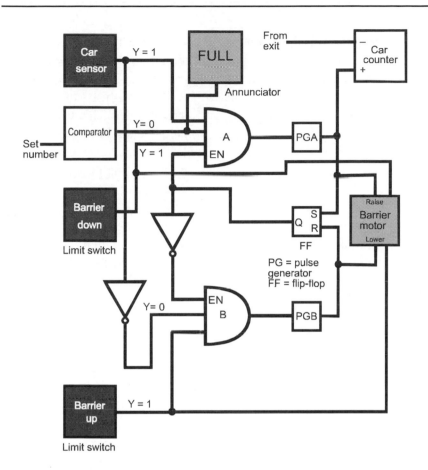

Figure 4.4 *This circuit for controlling the entrance to a car park is built from logic ICs, ranging in complexity from simple AND and NOT gates, to a complex counter and magnitude comparator. The system comprises sensors (dark grey) and actuators (light grey) linked by logic elements (white).*

until the 'barrier up' limit switch detects that the gate is fully up. Then it stops in the raised position. At the same time the pulse increments the car counter circuit. This counter has a second input from the exit gate which decrements the count every time a car leaves the park. The pulse also goes to a flip-flop which changes state, so ending Phase A.

With gate A now disabled and gate B enabled, the system enters Phase B. With the barrier up (high output from limit switch) and the car still at the barrier, nothing happens. As soon as the car has driven forward and cleared the sensor, all inputs to gate B are high. Its output goes high, and a pulse is generated by pulse generator B. This triggers the barrier motor to lower the barrier, which it does until the 'barrier down' limit switch gives a high output. The pulse also triggers the flip-flop to change state, ending Phase B and putting the system back into Phase A ready for the arrival of the next car.

When the number of cars in the park equals the preset 'full' number, the magnitude comparator gives a 0 output, which causes the annunciator lamp of the 'FULL' sign to be switched on. The barrier cannot be raised while this is on.

This system includes many of the commonly used elements. There are sensors (optical sensor, limit switches), and there are actuators (barrier motor, annunciator). There are a number of different logical devices such as AND gates, inverters (NOT gates), a counter, a magnitude comparator, a flip-flop, and two pulse generators (or monostables). All of these are available ready-made as members of the TTL and CMOS logic families.

The task of controlling the entrance barrier appears to be a simple one, but the logic of Fig. 4.4 is already looking highly complex. The circuit for driving the barrier motor needs additional logic, not shown in the figure, to interlock the operation of the motor with the signals from the limit switches.

Although we have tried a dry run on paper, it really needs to be built and tested, perhaps as a model or on a simulator, to make sure that it works. Even at this stage, a number of flaws can be seen. For instance, how does the system distinguish between a car waiting at the barrier and a person who happens to pass by and break the beam of the optical sensor? Another question we might ask is what happens if the driver decides not to park after all and backs out without going under the barrier? Providing for these and other contingencies will make the logic more complicated. There are also features that we might like to add to the system, such as a machine to issue a ticket at the barrier and to wait (delaying the raising of the barrier) until the driver has taken the ticket. We should also allow for manual operation of the barrier in case of failure of the system or a power failure.

The important point is that the logic system is complex and is hard-wired. If we discover faults in the system or want to add improvements after the system has been set up, it is likely to require the addition of several more ICs to the system board and several more tracks. It may be simpler to scrap the existing board and build a new one. Hard wiring provides a system that works (if previously tested and found to be faultless), and one that has no program that can possibly become corrupted. But against this is the inflexibility of the system once it has been designed and installed.

Test yourself

1 An analogue signal is one that:

        A repesents a sound wave.
        B comes from a sensor.
        C takes any value within a given range.
        D is either high or low.

2 The precision of a typical resistor is:

        A 1%.
        B 10%.
        C 0.01%.
        D 0.1%.

3 The precision of a 12-bit binary value is:

        A 1%.
        B 0.2%.
        C 2%.
        D 0.02%.

4 A well designed digital circuit may be affected by:

        A temperature.
        B quantisation noise.
        C accuracy of components.
        D electrical interference.

5 In relay logic, the logical AND function is produced by:

        A two normally-open relays in parallel.
        B two normally-open relays in series.
        C a normally-open relay and a normally-closed relay.
        D two normally-closed relays in parallel.

6 A TTL logic gate is faster than a typical relay by a factor of about:

        A 10.
        B 100.
        C 1000.
        D 1000000.

7 The choice below that is not a logic family is:

        A TTL.
        B EMI.
        C CMOS.
        D ECL.

5
Programmable controllers

The controllers described in the previous chapter are fixed function controllers. They are built to perform one or a few functions and, if they are well designed, perform them reliably for many years. Some of them can be switched to perform a range of related functions. For example, a clothes washing machine can be set to perform a limited range of washing and rinsing operations. Problems arise when we want to extend or alter the functions that the system can perform. The systems are hard-wired and the wiring must usually be altered. It will probably be necessary to change some of the ICs and perhaps add extra ones. Changing the basic action of the system is likely to be unduly expensive.

The systems described in this chapter depend on software rather than hardware. In other words, they are *programmable*. Programmable systems are of three main kinds:

- Programmable logic controllers.
- Microcomputer systems.
- Microcontroller systems.

In this chapter we examine each kind of controller in turn, pointing out the features, capabilities and applications of each.

Programmable logic controllers

A programmable logic controller, or PLC, is probably the control device most often used in industry. A PLC is essentially a microcomputer, based on a microprocessor with a system clock, ROM, and RAM. In general, the microprocessor is simpler than that found in a PC or other desktop microcomputer. Its instruction set is limited to fairly basic operations. The supporting memory is much smaller than that found in a typical microcomputer, there might be only a few control buttons or keys instead of a full 104-key keyboard. A PLC is unlikely to have any sort of disk drive. Its display is usually of low resolution and may consist simply of a single line of LCD characters. It may have a printer attached to it, but this is likely to be only a basic thermal printer capable of printing price labels or similar elementary matter.

So much for what a PLC *does not* have. What is *does* have is a large number of input and output ports. The input ports receive data and instructions from sensors, such as buttons, keys, limit switches, temperature sensors, light sensors, and strain gauges. Many of the input ports receive only a single bit of data. In other words, the input line is at logic high (1) or logic low (0). However, some inputs receive multi-bit data, or analogue data. Similarly many of the output ports provide only a single bit for turning a lamp on or off (for example) or for controlling a solenoid valve. Others may have multi-bit outputs, for example, for controlling a stepper motor.

Fig. 5.1 shows the layout of a typical PLC system. It is normally housed in a cabinet, with the control keys and indicator lamps mounted on the door of the cabinet. The units of the system are supported on racks inside the cabinet. As well as the control unit (the simple microcomputer), there will be a low-voltage power supply for the system, and a number (sometimes a large number) of *cards*. The purpose of the cards is to interface the control unit to the sensors (input cards) and actuators (output cards). Each sensor has its own card, identified by the number of the rack it is in and its position within the rack. Similarly, each actuator has its own identified card. One of

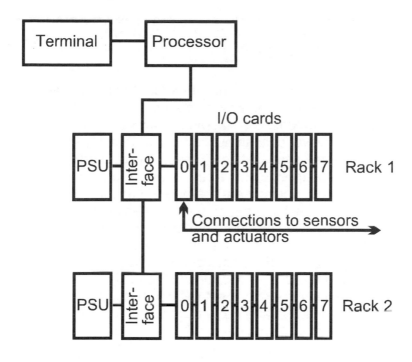

Figure 5.1 *A PLC control system is built up from a range of ready-made units that can be connected together in many different configurations, as required. This flexibility is one of the reasons why PLCs are so widely used for control.*

the advantageous features of PLCs is that the engineer can select an assortment of cards to suit the various sensors and actuators in the control system. It is also possible to include some spare cards that are unused but which are available for future developments in the system. Because the system is built up from a number of interchangeable units, it is easy to replace the existing cards with different ones if the system design is updated.

The cards are designed to operate with DC or AC inputs or outputs. The circuitry on a DC card includes transistor switches to sense the logical state of the input signal or to switch an external source of power to an actuator. The circuits always incorporate devices such as

PLCs for packing

A small PLC system is sufficient to control a machine that stacks a batch of wooden spindles ready for packing in a cardboard box. As the spindles arrive at the machine on the conveyor belt, an optical sensor that has its beam directed across the belt counts them. The spindles are lined up side by side until a single layer of them has accumulated. Again, there are optical sensors to detect when the correct number of spindles has been accumulated side by side. Then the next layer is built up on top of the first one. More layers are built, occasionally interspersed with thin strips of wood laid across the pile to accommodate any variations in the thickness of the layers. Such variations, if left unabsorbed, could make the stack unstable and liable to collapse. The stack is built up layer by layer until another optical sensor tells the controller that there are enough layers. Then the whole stack is moved away to be boxed. All of the actuators in this machine are electro-pneumatic. Microswitches are used as limit switches to detect when the actuators have reached their correct operating positions.

optoisolators to isolate the logic circuits of the controller, usually running at 5 V DC, from the external circuits which may operate on a range of DC voltages up to 50 V. The cards are also designed to isolate the computer circuitry from large power surges, often produced in industrial plant when inductive loads are switched. AC input cards have a rectifier to convert the incoming AC signal to DC. AC output cards may have a triac circuit for controlling motors and similar actuators. All cards generally have LEDs to indicate the logic state of input or output, so making it easier for engineers to monitor the operation of the system.

Keeping up?

1 List items that are usually part of a computer system but are generally missing from a PLC system.

2 List the main features of a PLC system.

3 Describe the main kinds of input and output card in a PLC system.

Programming a PLC

Another feature that has made PLCs widely used is that programming is relatively simple. This is because the 'program' is usually written in a form that looks very much like the familiar and easily understood diagrams of relay logic systems. This appeals to engineers who are familiar with the older technology as well as being easily learnt by those who are not. The idea is that it should not be necessary to employ a specialist computer programmer to set up a PLC system. The fact that PLCs can perform only a limited range of operations makes it possible to employ this programming technique.

Fig. 5.2 shows a PLC 'program' which has the same function as the relay logic system of Fig. 4.3. With the control unit in programming mode, the diagram is entered on the screen. The keyboard of the control unit has special keys for each of the relay logic symbols. The program 'lines' are numbered (the example shown is line 46) and the symbols are each identified with a number or code which indicates the input or output card. The system used on Mitsubishi PLCs uses the 'X' prefix for inputs and the 'Y' prefix for outputs, but we have allocated the card numbers fortuitously in this example. Annotation on the diagram helps the user to understand how the system functions.

In general, a PLC program of this type is easy to understand. It can be seen that, when two or more symbols are drawn 'in series', this represents the AND operation. So the program of Fig. 5.2 reads:

If X0 AND X13, then Y30

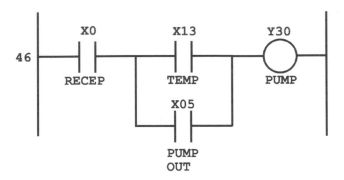

Figure 5.2 *The program of a PLC performing the same function as the relay logic of Fig. 4.3 has much the same appearance as the symbolic relay logic circuit. Symbols marked 'X' are inputs, and those marked 'Y' are outputs. The numbers refer to the cards that are wired to the corresponding sensors and actuator.*

Control systems at work – 5

Opening this cabinet beside a colour printing press reveals three PLCs (right). Each has a keyboard of a few keys (it is programmed by downloading the program from a PC) and a simple  alphanumeric display. Ranged along the upper and lower edges of each front panel are numerous input and output terminals. Some of these are wired to the relays and other interfaces mounted on the left side of the cabinet.

In words, if the receptacle is in position and the temperature is high enough, the pump is switched on. The program has a second symbol X05 in parallel with X13. Symbols in parallel imply the OR operation for there are two ways in which 'current' can flow through from X0 to Y30. So the whole program reads:

If X0 AND (X13 OR X05) then Y30

The part in brackets must be considered first, before the remainder of the statement. In words, the meaning is that if the temperature is high enough or the manual switch is pressed, and at the same time the receptacle is in place, then the pump is started.

Fig. 5.3 sets out the program for controlling the entrance to a car park. It performs the same function as the logic circuit of Fig. 4.4, except that it does not count the number of cars in the park and does not refuse entry when the park is full. Additional programming could perform these functions. The program illustrates a few additional points about PLC programs. In the first line of the excerpt shown here, if an optical sensor detects that a car is waiting to enter, and the barrier is down (as indicated by a limit switch) and the program is in Phase A, a memory location in the controller is set to a '1'. The use of memory is something we have not described before. A substantial area of the memory of the PLC is set aside as 1-bit 'flags'. These can be set to hold '1' or reset to hold '0'. In the program, the memory locations can be inputs or outputs. That is to say the PLC can read data from the memory location, or can write data into it. A number with the prefix 'M' often identifies memory locations. In the first line of Fig. 5.3, the location M20 is set to 1 when Phase A is in operation. If this is so, then a 1 is stored in location M41 to indicate that it is permissible to raise the barrier.

The next line is line 70. Note that the lines need not be numbered consecutively. This makes it easy to interpose additional lines, if necessary, as the program is developed. In line 70, if Phase A is operative and the barrier is up, this marks the end of Phase A and the

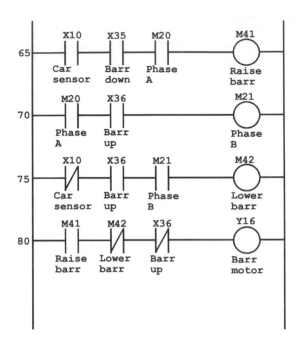

Figure 5.3 *The 'program' that corresponds to the part of the logic of Fig. 4.3 concerned with raising and lowering the barrier of the car park. It has the appearance of a ladder with four rungs, which is why PLC programs set out in this way are described as 'ladder logic'.*

beginning of Phase B. M21 is set to indicate this. In line 75, the first symbol inverts the action of the signal from the sensor by using a relay with normally closed contacts, which are open if the car is there. If the car is not at the barrier, and the barrier is up and it is Phase B, then the flag M42 is set to allow the barrier to be lowered. In line 80, if the flag to raise the barrier is 1, and the flag to lower the barrier is 0, and the barrier is not up, the barrier motor is started (a Y output).

This simple segment of a program does not show how the flags are reset and does not show how we cause the motor to turn in the direction that raises the barrier. The complete program needs many more steps. But Fig. 5.3 does illustrate how memory locations may be

Programming languages

Machine code: The binary code groups actually recognised by the microprocessor are stored in consecutive locations in memory. They represent instructions to the microprocessor (op codes) and data for the microprocessor to use. When the program is run it is read into the microprocessor one group at a time, in order, and then executed. Op codes, written in hexadecimal (for example, CE 04 A5) bear no recognisable relationship to the operations they are coding, which makes it very difficult for a programmer to work directly in machine code.

Assembler: This is a program by which the programmer can type instructions in the form of mnemonics. These are abbreviations that indicate the nature of the operation being coded. For example 'LDX 5C', might mean ' load register X with the value 5C'. When the program is finished, the assembler converts it into machine code.

High level language: Such a program allows the programmer to type instructions in a form that approximates to ordinary English. For example, in the popular high level language known as BASIC, it is possible to type a command such as 'If weight > 500 then print "Overweight" else print "Within limit" '. Such a command requires several dozen machine code groups to execute it, but the BASIC program is able to generate these instructions automatically.

Not all high level languages are as easy to use as BASIC but all are easier and quicker to use than either assembler or machine code.

used as flags to register the current state of the system, and how these locations may be used several times within a program. It shows how a location used as an output on one line may be used as an input on others. It illustrates the use of normally closed 'relays'. Finally, it illustrates why this type of program is known as *ladder logic* and why the stages in the program are known as *rungs*.

The way a ladder logic program is executed differs in an important way from the program of a microcomputer. For example, when a BASIC program, is run, the computer starts at the first line, and executes it. Then it goes to the second line and executes that, and so on until it comes to the end of the program. It may sometimes branch to other parts of the program when it encounters a GOTO command or when it is running in a FOR ... NEXT loop, but the essential point is that it executes the program line by line. On any line, it may receive input, from an external device or by reading a value in memory. On any line, it may generate output, perhaps by sending text code to a printer.

With ladder logic, the program scans through the whole program over and over again in a continuous loop, taking about 0.1 s for each scan. Before it scans the program, it reads in the state of *all* the system's input devices. The state of each sensor is stored in memory and the stored data remains fixed while the PLC runs through the program. At each line, the sensor inputs are taken to be as they were just before the scan. Each line processes the information present at the beginning of the scan and produces an appropriate output. However, this output is not acted on yet. It is acted on only when the scan is finished. Then solenoids are energised or de-energised, motors switched on or off, and indicator lamps are updated. The PLC then reads new information from the sensors and the scan begins again.

The scanning system means that there is an inevitable delay between a change of sensor input and the corresponding change in actuator output. The delay is unimportant in most industrial systems. It can be minimised by careful programming. For example, line 65 sets the flag M41 and then line 80 uses the setting of M41 to enable the motor. The motor is not actually turned on until the end of the scan. If lines 65 and 80 had been placed in the reverse order, M41 would be set during one scan but its effect could not occur until the end of the next scan.

The more elaborate PLC systems include a terminal for use in programming the system, but not all systems have this. Many of the simpler systems have a hand-held programmer, which is connected to the control unit to program it or amend the existing program. This is economic, for a single programmer unit can be shared between several PLCs. It is also possible to program some models of PLC by connecting it to a microcomputer that is running special programming software. One of the advantages of PLCs is that programs can be run and tested off site. Often an input simulator is used, which consists of a bank of switches to produce high or low inputs corresponding to the input normally available from the sensors of the actual system.

Although ladder logic is widely used for programming PLCs, some of them can also be programmed in other ways. A few are programmed by drawing on a screen a representation of a *logic circuit*, using symbols for conventional AND, OR and other logic gates. The drawing may also include other devices such as counters, timers and shift registers. When the drawing is complete, the programming computer works out the logical equivalent of the drawing and compiles this into code that is understandable by the PLC.

Another programming technique, much more akin to conventional programming, is the use of a *statement list*. This consists of a series of instructions to the PLC coded in a language that is reasonably similar to English. An example of this is the programming of the SPLat controllers. SPLat is short for 'Simply Programmable Logic Automation Tools'. Although the controllers can respond to a wide range of instructions, the manufacturers consider that most industrial programming can be done by using only a very small instruction set.

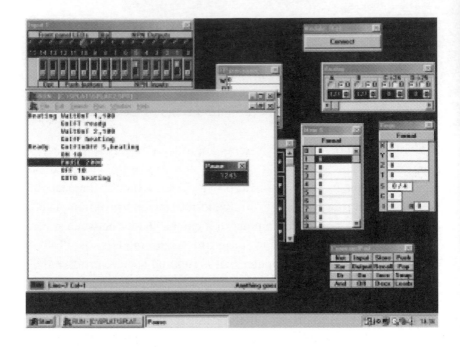

Figure 5.4 *PLCs are programmed by using a special programming computer or by running special software on an ordinary computer. Here, a program for a SPLat PLC is being tested on a PC before being downloaded into the PLC.*

These they call the Fast-track Instructions. There are just 14 of these and, since some of them are simply the opposites of others, there is very little to learn.

The system has purely digital input and output, so everything can be handled with 0s and 1s. Fig. 5.4 shows the screen of a PC when it is running the SPLat programming software. At top left is a display that simulates the input and output panels of the PLC. The SPLat has 8 npn transistor inputs, 5 push buttons and 2 optical sensors. It has 8 npn transistor outputs, 7 front-panel indicator LEDs, and an output beeper.

An editing window in Fig. 5.4 displays the program that is currently being developed. The listing there corresponds to the logical circuit of Fig. 4.1, and is as follows:

```
Heating    WaitOnT 1,100
           GoIfT Ready
           WaitOnT 2,100
           GoIfF Heating
Ready      GoIfInOff 5,Heating
           ON 10
           PAUSE 2000
           OFF 10
           GOTO Heating
```

Input 1 is from the temperature sensor and goes high when the correct temperature is reached. Input 2 is from the manual switch and goes high when this is pressed. Input 5 is from the proximity detector, which goes high when the receptacle is in position. Output 10 controls the valve. 'Heating' and 'Ready' are labels. The program first waits for Input 1 to come on (= temperature correct). If it is on already or if it comes on within 1 second (100×10 ms), the program drops through to the next line where it is told to go to Ready if the previous statement is true (which it is). If Input 1 does not come on within 1 second there is a timeout and the program drops past the next line to be told to wait for the manual switch to come on within 1 s. If this does not come on the next line (Go if false) sends it back to heating to begin again. If temperature is correct or the manual button is pressed, the program reaches the Ready label. Where it is sent back to the beginning if input 5 is off (receptacle absent). Otherwise, it runs the rest of the program, turning the valve on, waiting 20 seconds for the liquid to drain, turning the valve off, and then returning to the beginning of the program.

The program is running in Fig. 5.4 with the currently processed line, the PAUSE command, being highlighted. A small window is counting down the period of the pause. The panel simulation shows switches 1 and 5 pushed up (= 1) and LED 10 turned on. Other windows on the screen relate to other aspects of programming the PLC.

This account covers only the essential features of PLCs and their programming. PLCs can perform much more complex tasks such as handling analogue data and performing calculations with it. Operations such as square roots and trigonometrical functions are available. Programs can have subroutines, called when the need arises. It is also possible to program the PLC for 3-term (PID) control, keying in the set point, the gains and the time constants required.

PLCs are used for controlling complex industrial plant, and other systems of comparable complexity. At the other extreme, there are very simple PLCs such as the Siemens LOGO! and the Matsushita FP0. The FP0 has only 6 inputs and 4 relay-type outputs. These smaller PLCs are intended for controlling very small systems such as those for automatically opening and closing a shop door or controlling a single motor. The system is housed in a single enclosure, smaller than a videocassette, with a control panel and display on the front and all the interfacing built in. The engineer simply has to provide a power supply and connect the inputs and output directly to the terminals on the I/O panel.

Keeping up?

4 What is ladder logic? How is it related to relay logic?

5 Write a short ladder logic program to control the opening of an 'up and over' garage door.

6 What are the differences between the form and execution of a ladder logic program and those of a statement list.

Microcomputers

Microcomputers are often used in control systems when:

* The system requires complex calculations or logic to be performed in real time.

- The system is being developed or is experimental in nature and constantly needs to be re-programmed.

- It is necessary to have input from a full 104-key keyboard.

- The system must provide a high quality graphics output, either on a monitor or a printer.

The microprocessor is at the centre of any computing system, whether it is a PLC or a microcomputer. The difference is that the microprocessor in a microcomputer is generally a much more advanced device. At the time of writing, the most commonly used microprocessor is a member of the Intel 8086 family, the most recent member of which is the Pentium III. This is intended to operate much faster and to perform a greater range of operations than the microprocessor of a typical PLC.

Fig. 5.5 shows the main parts of a typical microcomputer system and how they are interconnected. The microprocessor is the central processing unit, or CPU. It is the controlling centre of the system and responsible for most of the activity that takes place.

The CPU is linked to all other parts of the system by three sets of conductors. Each set, or bus, consists of a number of conductors running in parallel with each other.

The address bus usually has the most conductors. It may consist of 20 or more conductors. When a given part of the system (for example a location in memory) is to be addressed the CPU places a low (0 V) or high (3.3 V or 5 V) voltage on each line to form the equivalent of the address (in binary) of the location. Note that, in Fig. 5.5, the arrowhead on the address bus (black) points away from the CPU indicating that this is only one-way communication.

The data bus (medium grey) has fewer conductors, often 8 or 16, and is used for sending data from the CPU to other parts of the system, or in the reverse direction. The arrowheads indicate the two-way communication.

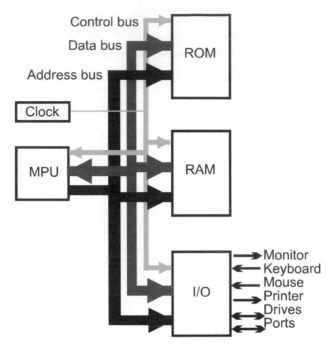

Figure 5.5 *A block diagram of a typical microcomputer. The heart of the system is the microprocessor (MPU) connected to all other parts of the system by the three busses for addresses, data and control.*

Finally, there is the control bus, which sends control signals between various parts of the system. This is a two-way bus, and includes a system clock line by which the timing of the operation of the CPU is controlled and synchronised with that of other parts of the system. This is a square-wave oscillator with a frequency (nowadays) of several hundred megahertz and is a necessary part of the system. It is not to be confused with a real time clock, which is an IC that keeps track of the time of day, and the date and is a useful but not essential item.

Two of the blocks in Fig. 5.5 represent the memory of the microcomputer. This is essentially of two kinds. ROM, or read-only memory, is permanent and mainly holds the programs needed to get the CPU started when it is first switched on. RAM, or random access memory, is temporary memory. Any data stored there is lost when the computer is switched off. RAM is mainly used for holding the program that the computer is presently running and any data associated with it.

Aiming a telescope

When the Lovell Telescope (then the Mark 1 radio telescope) at Jodrell Bank began operations in 1957, its control was purely analogue, but it has been digital since 1970. When the telescope is being moved a long distance it is accelerated at its maximum safe rate and then driven at maximum safe angular velocity (see Fig. 2.11). For the Lovell telescope, the maximum velocity is 9°/min in azimuth (horizontal rotation) and 6°/min in elevation. Smaller telescopes may be moved faster. Typical rates are up to 40°/min in azimuth and up to 10°/min in elevation.

When the telescope has reached its maximum angular velocities, these are held constant until the telescope nears its final position. Then, when the telescope is pointing within 5 or 6 degrees of its destination, it is decelerated so as to come to rest at exactly the desired angles. This is all under computer control. Sensors (Ch. 3) monitor the position of the telescope and corrections are applied to allow for refraction of radio signals in the Earth's atmosphere. There are also corrections for the mechanical distortion of the framework and dish under the load of their great weight. When all of these factors have been allowed for, digital signals are sent to control centres on the telescope. These in turn send analogue voltages to the motor controllers. Once on target, the telescope is moved very slowly to track the celestial object, the computer calculating the rate of change of azimuth and elevation required.

A microcomputer has a much larger capacity for storage than a PLC. A typical microcomputer has a large RAM of 64 Mbytes or more for holding programs and data at run time. It also has at least one hard drive, with a capacity rated in gigabytes, and usually has a CD-ROM drive and floppy disk drive as well. These are needed because a microcomputer is intended for running many different programs, most of which require storage rated in megabytes. It may be required to run several such programs simultaneously. For example, the author often runs a word processor, a graphics drawing package, and an electronic circuit simulator all at the same time. By contrast, a PLC runs only the one program for controlling the machine to which it is connected. The same program runs day in and day out, for months or possibly years. The number of program steps is reckoned in hundreds and a data storage memory of 2000 registers is considered to be ample for most applications.

The drives are part of the input/output (I/O) section of the microcomputer. Other parts are the monitor, the keyboard, possibly a mouse and a printer. Another major difference between a PLC and a microcomputer is that the former can be provided with a large number of input and output cards (Fig. 5.1). This is an essential feature of a controller that has to receive input for perhaps several dozen sensors and provide output to a similar number of actuators. Moreover, the input and output cards are designed to be very robust for operating reliably under difficult conditions on the factory floor. A microcomputer does not usually have input and output on such a scale. Usually there are a number of edge-connector 'slots' on the motherboard of the computer. Eight slots are generally thought to be an ample number. Cards are plugged into these slots to connect them to the data, address, and control busses of the computer. If more cards are required than can be installed on the motherboard, the busses can be extended through buffers to an external rack that can accommodate many more cards (see box).

A typical computer usually has several commercially built standard cards permanently installed in its slots. These include special high-resolution graphics cards for interfacing to the monitor, and sound cards for interfacing to audio or MIDI systems. Printers, plotters and

Control systems at work – 6

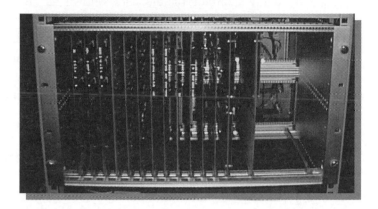

This rack holds the circuit cards that interface the DEC MicroVAC 2 computer at Jodrell Bank radio astronomy observatory to the sensors and actuators of the Lovell Telescope. The cards are each edged with a row of LEDs to display the current state of the logic circuits. The long rows of LEDs on the two middle cards indicate in binary form the present azimuth and elevation of the telescope. These LEDs can be seen to flash very rapidly in binary sequences as the telescope is homed on to a celestial object.

scanners may also have their own cards. General-purpose I/O cards are available, but for one-off tasks it is likely that the cards will be custom-built to suit the requirements of the system.

Whether a card is commercially built or homemade, there are certain essential circuits that it must include. The most essential of these is address decoding. All cards are connected to the computer busses so there must be some way of activating a card so that it can put data on to the data bus or receive data from it. When a card is to be contacted, the microprocessor puts the address of the card on the address bus; and then reads or writes data. The address that has been placed on the bus is recognised by a logic circuit on the card known as an *address decoder*. A card can be allocated to only one address, but very often, it may cover several adjacent addresses, using each of these for different but related functions. When the decoder recognises an address on the bus as being its own address it immediately enables the memory or other chips on the card. At other times, these chips are disabled and can neither send nor receive data.

Cards used in control systems may carry special circuitry to accept DC or rectified AC input from external devices. The cards are normally designed to interface the standard 5 V of most logic systems to the higher voltages such as 12 V and 24 V that are more often used in

Testing a gearbox

This rig was designed to test a gearbox that is part of a flight control system (Fig. 5.6). A motor (Motor 2) normally drives the gearbox and the driven shaft carries an aileron or spoiler. In the test rig, a pair of motors (1 and 3) simulates the opposing force due to the airstream over the flight surfaces. There are sensors to measure the torque produced by the motors, and there are encoders to detect the angular position of each motor. There are also sensors on the test bench to measure the temperature of the motors and the extent of mechanical vibration of the system.

The parts of the system are linked by various means, including an RS422 serial digital link, and a number of 1-bit digital links. The 1-bit signals are mainly for emergency use, for example, to turn off the motors instantly if the torque becomes excessive. Fibre-optics carry the signals from the shaft encoders to the control computer, and are immune to electromagnetic interference from the motors.

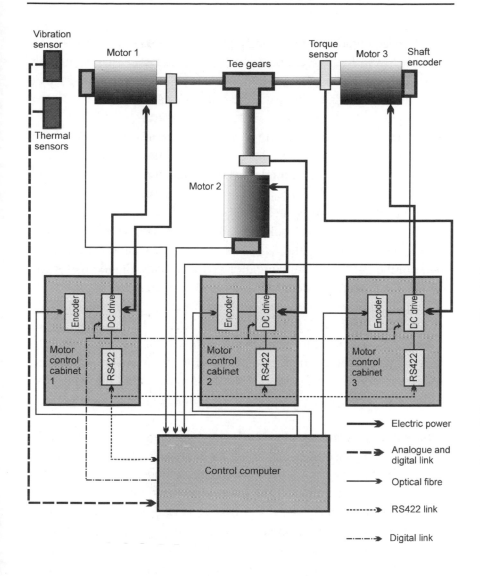

Figure 5.6 *A computer-controlled test rig for a flight control gearbox.*

industrial equipment. They usually have opto-isolators to protect the computer from high voltages that may be generated when switching heavy inductive loads. It is common for such cards to have a row of LEDs at the rear edge to indicate the state of each channel.

For small systems operating in favourable conditions, the microcomputer may be a desktop or laptop model using interface cards in the slots on the motherboard or in an external rack, connected to an extension of the busses of the computer. For use in industrial environments, the whole system can be built up on a rack with all units connected to a common backplane bus. There are several standard busses, one of which is the STE bus. This is very similar to the bus found in the IBM personal computer and its clones. Standards for the STE bus are defined in detail so that a wide range of processor boards (PC-type computer on a single board) and interface cards are available off the shelf for the engineer who has the task of setting up a control system. Any of this wide range may be chosen and incorporated into the system with the confidence that it will work. A system assembled from these units is very robust both electronically and mechanically, suiting it for conditions encountered on almost all industrial sites.

Robustness is a feature of the hardware of computer systems used in industry. It may also be a feature of the software. This is particularly the case in the avionics industry where safety issues dominate system design and construction. One of the factors holding back the application of computers has been that software occasionally proves unreliable. The average computer user is familiar with the occasional error message appearing on the screen: "Exception error: save your data and close the program". This is not something to be done when a computer is controlling the flight of a jet plane a mile high above the Pacific. For this reason, we find that although popular languages such as FORTRAN, PASCAL and C may often be used in control systems, there are many systems in which other languages are preferred. One of these is ADA, which is a difficult language to learn but which is very robust in action. Special languages have been developed with a set of commands appropriate to a certain type of control system. An example is LUCOL™, a proprietary language used by Lucas Aerospace. Ainstructions, each consisting of 14 bits, and it has 36 bytes of RAM. It also has a built-in timer. Some other members of the family have an analogue-to-digital converter or other useful devices such as an analogue comparator or an oscillator. Other members of the family and microcontrollers of other families, such as the Atmel Mega, have rather more extensive facilities, such as snother is FADEC, developed by Lucas in conjunction with Rolls Royce for engine control. Its full name is Full Authority Digital Engine Control.

Languages may also help speed the work of system designers. One such language is HP VEE. This is a visual language, one in which the programmer can assemble functional blocks on the screen, representing the parts of the control system and plant. They can be linked together on screen by drawing lines to join them. Their parameters are typed in as work proceeds. Then the whole system can be simulated on screen to check that it works as required. Finally, the program generates the code for the processor. Working with this software is much easier than using a conventional high-level language such as C. It is also much easier for other people to examine the program and understand how it works.

Keeping up?

7 Under what circumstances is a microcomputer preferable to a PLC?

8 In what ways can a microcomputer-controlled system be safeguarded against the effects of breakdown?

9 What kinds of circuit would you expect to find on a card interfacing a computer to a steel rolling mill?

Microcontrollers

A microcontroller is a microcomputer on a single silicon chip.

Different types of microcontroller vary in their contents but a typical microcontroller comprises a CPU, a system clock, a block of ROM, a block of RAM with a number of I/O terminals. An example is the PIC 16C61, one of the simpler and cheaper members of the extensive PIC family. This has a CPU, ROM, RAM, and 13 I/O pins. Its programmable erasable ROM can hold up to 1024 machine code everal kilobytes of ROM and RAM and more I/O lines. There is a wide range of microcontrollers to choose from and the choice depends on what features will be made use of in the proposed control system.

It follows from the description above that the CPU of the microcontroller does its job in the same way as in a microcomputer, but its access to memory is different. Before a microcontroller can be used, its ROM must be loaded with the program it is to run. It is intended that this program will remain in the ROM indefinitely and is run automatically from its beginning whenever power is applied or the reset button is pressed. The usual procedure is to program the ROM, often by using software on a personal computer. The program is then tested and debugged in the PC. Next, the tested program stored in the PC is downloaded into the ROM, a process that takes only a few seconds. If there are still faults, the ROM can be quickly re-programmed as often as necessary. Once the ROM has been programmed satisfactorily the program remains there for several years, even when power is switched off. The amount of ROM available for storing the program is very small in comparison with the disk storage of a microcomputer but it is sufficient for many control applications. Similarly the amount of RAM is relatively small, but it is big enough to store data temporarily while the program is running.

Although microcontrollers are usually programmed by using an assembler, the BASIC Stamp®, manufactured by Parallax Inc. is programmed in BASIC. This makes it very quick to set up a control system. The Stamp, so named because its circuit board is little bigger than a postage stamp, consists of a PIC microcontroller together with a ROM that is already programmed with a BASIC compiler. The user types the BASIC program using software running on the PC. Then it is

A typical microcontroller

The Atmel AT90S1200 microcontroller is one of the simpler devices of the Atmel range, yet it displays a number of useful features. The pinout (Fig. 5.7) shows that 15 of its 20 pins can be set as inputs or outputs. These are organised into two ports, Port B with 8 bits, addressable as a single byte, and Port D with 7 bits. Two of the pins are used for the supply lines; the chip runs on any voltage in the range 2.7 V to 6 V. Two of the pins are for an external timing crystal; the system can run at up to 16 MHz. The remaining pin is used to reset the system. Three of the pins of Port B are also used as a serial download port.

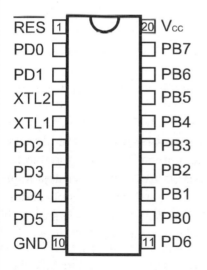

Figure 5.7 *The Atmel AT90S1200 microcontroller is enclosed in a 20-pin d.i.l. package of the type used for TTL and other logic ICs.*

As well as a CPU, and a control unit, the chip carries four blocks of memory: 1 kilobyte of flash ROM (for programs), 64 bytes of EEPOM, and 32 8-bit data registers. In addition there is an 8-bit timer, a watchdog timer and an analogue comparator.

downloaded into the Stamp, where it is converted into machine code and loaded into the ROM of the Stamp. From then on, the Stamp may be removed from the programming board and inserted into a socket on the circuit board of the control system.

Distributed processing

More recently, new types of sensor and actuator incorporate a microcontroller to supervise their operation. This leads to the concept of *distributed control systems*. This is the basis of sophisticated flight control in certain aeroplanes. An example of this is the 'fly by wire' system developed for use in the A320 Airbus and also the A330. The system as a whole is under the control of a computer, which is on the flight deck and is ultimately under the control of the pilot. The control of individual flight surfaces such as an aileron or the rudder is controlled locally (for example, within the wings) by a microcontroller built in to the actuators. The main control computer issues an instruction to the actuator to move the aileron to a given angle. This instruction is received by the microcontroller, which then proceeds to execute it. It sends signals to the motor responsible for moving the aileron and at the same time receives signals from sensors, reporting the exact position of the aileron instant-by-instant. The actuator accelerates the aileron at the maximum permissible rate, then moves it in the appropriate direction at the maximum permissible rate, taking account of resistance experienced due to airflow over the wing. As the aileron nears its intended position the rate of tuning is decreased, so as to bring it to rest at exactly the required position, without overshooting. This can be very important in a structure such as an aileron, for overshooting may force the aileron down against the wing surface, damaging both the aileron and the actuator mechanism. However, in all flight control systems it is essential that a fault (in this case a failure to halt the aileron in time) can not result in disaster. Aileron control tests show that there is no material that is suitable to cushion the aileron against the wing. The only solution is to include a slip clutch in the drive.

At all stages in the movement of the surface the microcontroller within the actuator reports back to the main computer. It tells the computer how far the flight surface has been moved. Eventually it tells the main computer that its task has been completed as required. If the microcontroller experiences difficulties in carrying out its task, it reports back to the main computer, which responds accordingly.

Mention of system faults reminds us that the hardware and the software of computers are extremely complicated and that therefore there are many ways in which they can eventually go wrong. We have already mentioned the use of specially devised proprietary software to counter this. Another approach is to provide redundancy in the system so that, if one element fails, there is a second or even a third identical element to take its place. Often there are two or three 'main' computers on the flight deck of an aeroplane or a system may include two or three identical microcontrollers, working in parallel. If the main units fail, one of the standby computers is automatically switched in. The same applies with sensors. An important variable is measured not by one but by two or three sensors. If there are three sensors and one disagrees with the other two, a warning is issued that something is wrong and an engineer is called to investigate. A common technique is to measure a quantity using two or more sensors that depend for their action on different principles. For example, in the control of flight surfaces the rotational speed of the actuator motor is often measured by using an optical shaft encoder (a rotary version of Fig. 3.5). At the same time, the speed of the motor is gauged by measuring its back e.m.f. The two measurements should agree. A warning is issued if they do not.

Distributed processing is now so well established that control systems can be built up from readily available standard cards. This is a modular approach and is known in the aircraft industry as Integrated Modular Avionics (IMA). Servicing and repair is greatly accelerated, because a faulty board can be replaced instantly. The modular approach makes a great contribution to safety. A more recent development is the Generic Smart Actuator Controller. This module is designed to be able to perform several different control functions. When plugged into the appropriate slot in the computer, it automatically reads the configuration of the slot and identifies the function it is intended to perform. From then on, it performs the function appropriate to that slot. Using smart controllers means that it is not necessary to stock a large range of different controllers, and removes the need to stock an extensive range of spares.

Because of their convenience and low cost, microcontrollers are being used more and more in control systems. In the home, one may expect to find a microcontroller in the washing machine, the dishwasher, the fax machine, and even in the 'smart' fan heater. The newer models of car have a microcontroller in charge of engine management. They are used in all kinds of scientific and measuring equipment. Their applications in all fields of electronics are increasing daily with the result that the most recent catalogues from electronics components retailers list several hundred different microcontrollers, each with its own special combination of CPU, RAM, ROM, timers, converters and other on-chip devices.

Keeping up?

10 What are the main differences between a microprocessor and a microcontroller?

11 In what ways are a microprocessor and a microcontroller alike?

12 What is distributed processing?

Test yourself

1 A PLC system has:

     A a floppy disk drive.
     B input cards.
     C a colour printer.
     D a joystick.

2 A mains powered electric fan is interfaced to a PLC through:

    A an AC output card.
    B an AC input card.
    C a DC input card.
    D a DC output card.

3 An opto-isolator on an input card is used to:

    A sense the logical state of the input signal.
    B display the logic state of one of the gates.
    C protect the computer circuit.
    D show when the card is receiving input.

4 A ladder logic program is similar to:

    A a BASIC program.
    B a relay system.
    C a logic IC.
    D a statement list.

5 In a ladder logic program, the OR function is produced by:

    A two normally-open 'relays' in parallel.
    B two normally-open 'relays' in series.
    C a normally-open 'relay' and a normally-closed 'relay'.
    D two normally-closed 'relays' in parallel.

6 One of the distinctive features of a PLC system is its:

    A high processing speed.
    B numerous inputs and outputs.
    C complex mathematical functions.
    D large-scale memory.

7 The main parts of a microcomputer system are linked by:

    A wires.   B buses.
    C cables.   D fibre-optics.

8 The system clock is used to:

    A make the computer run faster.
    B print the time on letters and other documents.
    C synchronise different parts of the system.
    D calculate the date.

9 Instructions passed along the data bus to the CPU are in:

  A assembler.
  B machine code.
  C BASIC.
  D ADA.

10 The instructions for a microcontroller are usually stored in:

  A an EEPROM.
  B static RAM.
  C a PC.
  D a CD.

6
Fuzzy control

Fuzzy control is a new way of developing control systems, based on *fuzzy logic*. This was invented in 1965 by Professor Lotfi Zadeh of the University of California at Berkeley. As has happened before when a radically new approach to science or mathematics has been proposed, the concept of fuzzy logic met with disbelief, disapproval, and often considerable opposition from the majority of the leading scientists and mathematicians in the United States of America as well as in Europe. These included both academic and industrial professionals, especially those occupied with research in Artificial Intelligence.

There were exceptions, such as Ebrahim Mamdani at Queen Mary College in the University of London, who devised a fuzzy logic system for controlling a steam generator. It had not previously been possible to control this generator adequately by conventional control systems. Hans Zimmermann at the RWTH, University of Aachen, used fuzzy logic for decision support systems. Fortunately, the themes of fuzzy logic were more acceptable to the outlook of scientists and mathematicians in Asia, particularly in Japan.

From the 1980s onward, Professor Zadeh's ideas stimulated much research into the theory and the applications of fuzzy logic. It was not long before practical and highly successful applications of fuzzy logic began to appear. Domestic appliances with fuzzy logic controllers have now been marketed by Japan for many years and have generally shown performance superior to those controlled in conventional ways.

The applications of fuzzy logic include pattern recognition and image processing, but most applications are in control systems. From the point of view of the public, the greatest impact has come from a host of 'smart' or 'intelligent' devices now commonly found in the home (Fig. 6.1). We shall look more closely at further examples later. Interestingly, there are also several applications of fuzzy logic to government, politics, and social action, but we shall confine our attention to control systems.

Crisp logic

When we are talking about 'ordinary' logic or, more precisely, Boolean logic and comparing it with fuzzy logic, we often refer to it as *crisp logic*. Crisp logic is the formal logic that has been used for centuries and today is the basis of the operation of most computers. The description 'crisp' alludes to the essential difference between crisp logic and fuzzy logic. Before we investigate this difference, we will examine a few important features of crisp, or Boolean, logic.

The main distinguishing feature of Boolean logic is that it is two-valued, or *binary*. In a Boolean world, everything is either true or not true (false). We conduct all our reasoning on this basis. The control systems that have been described so far in this book all operate on the same premiss. Take a thermostat as an example. Its action can be expressed as a logical rule of this form:

IF the temperature is below 15°C, THEN switch on the heating element.

Ignoring the slight complication of the hysteresis built into some thermostat circuits, this is just what a thermostat does. A wholly

Figure 6.1 *This 'Intelligent Fan Heater' warming the author's study is controlled by a microcontroller. When it is first switched on in a cold room, its fan whirrs gently, providing a warm flow of air, rather than a powerful and uncomfortable blast of cold air. In contrast, if the room has been warmed and the set point is reduced, the heater goes off and a copious flow of cold air reduces the room temperature as rapidly as possible. Heaters programmed in fuzzy logic act so as to produce accurate temperature control and the most comfortable environment with the minimum of noise and electric power consumption.*

mechanical thermostat (Fig. 1.12) expresses this rule in mechanical form. An electronic thermostat (Fig. 1.11) does it electronically. An equivalent statement in a program for the BASIC Stamp might be:

IF temp = 0 THEN heaton

Here *temp* is a variable which has the value 0 when the temperature is below 15°C and has the value 1 when the temperature is equal to or greater than 15°C. The label *heaton* makes the microcontroller branch to an output statement that switches the heater on.

At this point, we have illustrated several points about Boolean logic and about computers:

- Like Boolean logic, the operation of a computer is *binary*. Bits are 0 or 1, voltages low (0 V) or high (often +5 V), and, in our thermostat example, temperatures are either less than 15°C or they are NOT less than 15°C. The in-betweens do not exist or are meaningless.

- A computing program may consist mainly of *rules*.

- The rules are often *IF ... THEN statements*. They select between alternative actions, depending on whether the IF statement is true or false. This is a binary operation.

- The rules are applied *in series* (one after another) as the program runs.

Keeping up?

1 What is the essential difference between fuzzy logic and crisp logic?

2 Write a typical rule for a crisp logic program.

3 What can you say about the way the rules are applied in a crisp logic program.

Sets

The binary rules of Boolean logic give rise to *sets*. We can define a set by writing a list of its members. For example, the set A of room temperatures less than 15°C is defined by writing:

$$A = \{5, 6, 7, 8, 9, 10, 11, 12, 13, 14\}$$

This assumes that a room temperature will not be less than 5°C. It also assumes that temperatures are measured to the nearest whole degree.

Similarly, we can define the set B of temperatures equal to or greater than 15°C:

$$B = \{15, 16, 17, ..., 40\}$$

The maximum possible room temperature allowed for the purpose of this discussion is 40°C. Any given room temperature in the range 5°C to 40°C falls into one or the other set. There are no temperatures that occur in both sets. Such sets with clearly defined content are crisp sets. We can represent the set in a diagram, as in Fig. 6.2.

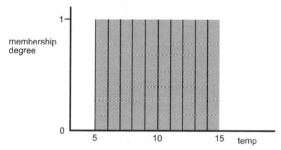

Figure 6.2 *The membership function of the set 'room temperatures less than 15°C' has a rectangular form. The vertical sides reveal that this is a crisp set.*

The figure shows that each of the integers 5 to 14 is a member of the set. Each of the members has a *membership degree* of 1. That is to say, it is *true* that it belongs to the set. All other temperatures in the figure have a membership degree of 0. It is *false* that they belong to the set. In other words, they do NOT belong to the set. The shaded area includes all the integer and non-integer values (if we decide to allow non-integers) in the set, from 5 up to, but not including, 15. The outline of the rectangular area defines the *membership function* of the set. The set is crisp, so its sides are vertical.

It is also possible to have a membership function with sides that are not vertical (Fig. 6.3). The figure represents a set of temperatures that might be described as 'comfortable'. What constitutes comfort is

The Sorites paradox

The paradox dates from the days of Ancient Greece and goes as follows.

Take a heap of sand. Remove one grain. Is it still a heap?

Remove another grain. Is it still a heap?

Continue removing the grains one at a time and each time ask the question: 'Is it still a heap?'.

Eventually, only one grain is left. Is this a heap?

Remove the grain so that nothing is left. Is this a heap? If not, when did it stop being a heap?

One way of solving the paradox is by specifying that a heap must consist of a minimum number of grains. Then take one grain away and it is no longer a heap. This is the binary Boolean approach that produces the absurd situation that, of two piles of sand, differing by only one grain, one is a heap and one is not. The fuzzy approach allows for all gradations between a large heap and nothing. This is a much more practicable way of thinking about heaps.

partly a matter of opinion. There are those who would consider temperatures in the upper twenties to be insufferably hot. These and others might feel that the low twenties is a little chilly. However, many people would accept 25°C being comfortable and the membership function shows this temperature to have a membership degree of 1. Below this temperature, the membership degree falls off. For example, 23°C has a membership degree of 0.6. This is something that is unthinkable in Boolean logic. In Boolean logic a temperature is either

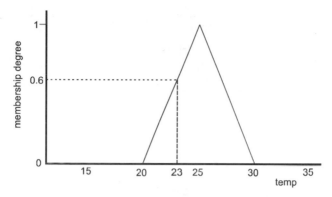

Figure 6.3 *The only member of this fuzzy set of comfortable temperatures with membership degree 1 is 25°C. Temperatures slightly higher or lower than this have lower membership degrees. For example, 23°C belongs to the set with a membership degree of 0.6.*

comfortable (1) or it is not (0). In the diagram, we rate its comfortableness as 0.6 on the scale of 0 to 1. This diagram describes a *fuzzy set*. The fuzzy set comprises a collection of temperatures ranging from 20°C to 30°C in which the members belong to the set to differing degrees. The further from 25°C, the smaller the membership degree. The set conforms to most people's concept of comfortable.

For obvious reasons, the membership function of Fig. 6.3 is described as *triangular*. This shape is often chosen for working with fuzzy sets but other shapes are sometimes used. The triangle need not be symmetrical, for instance. Fig. 6.4 illustrates a trapezoidal fuzzy set. The members in the middle range are all full members of the set. Those at the extremes have lower membership indices. In some applications, a function is outlined by quadratic and exponential curves. However, using a more complex function brings little benefit in relation to the complexity of the computations to which they give rise. The triangular and trapezoidal functions are the easiest to work with and are preferred for this reason.

It is possible to think of a series of fuzzy temperature sets spanning the range from Arctic cold to tropical heat, as in Fig. 6.5. Each set is given

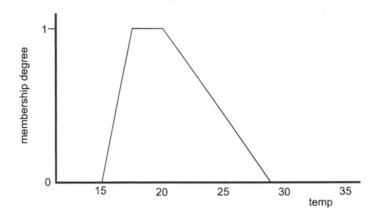

Figure 6.4 *A fuzzy set with a trapezoidal membership function might be termed 'temperate' by people who are prepared to put up with slightly warmer temperatures but who dislike the cold.*

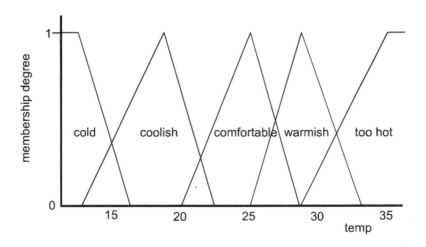

Figure 6.5 *Room temperatures may be divided into five fuzzy sets, each named by a linguistic variable.*

a name based on how the temperatures within each set seem to most people. This emphasises an important feature of fuzzy logic. Sets can be given descriptive names and these can be used as variables. Instead of defining temperatures as a series of precisely limited ranges, in degrees Celsius, we can use *linguistic variables*, such as:

Cold, coolish, comfortable, warmish, too hot.

This is typical of the way in which fuzzy sets are defined and named (Fig. 6.6).

This section has illustrated several features of fuzzy sets. Compare these with those listed for crisp sets on p. 136:

- Fuzzy logic is *multivalued*, not binary. Each member of a set has a *membership degree* (ranging from 0 to 1). There are many in-betweens and they are all meaningful. The pattern of membership degrees determines the *membership function* of the set.

- Fuzzy sets have *fuzzy boundaries,* which may overlap with adjacent fuzzy sets.

- Fuzzy sets are usually referred to by using *linguistic variables*.

As might be expected, fuzzy sets are the basis of *fuzzy rules but,* before going on to discuss how a fuzzy control system is built, we note the underlying differences between Boolean logic and fuzzy logic. The main one is that Boolean is binary and fuzzy is not. Boolean logic and digital computers have binary operation in common. This accounts for the dominance of Boolean logic in our ways of thinking. Yet there are many lines of reasoning that reach the same conclusion as the Sorites Paradox (see box).

Binary systems do not accord with many aspects of everyday life and common experience. It is rarely that things can be classified into black and white. There are nearly always in-betweens.

Artificial Intelligence

The idea of building a computer-like machine with intelligence equal to (or exceeding!) that of the human brain has inspired many researchers for many years. In spite of this, AI has produced little of consequence. One of the reasons for this is that AI is based firmly on Boolean logic. An AI computer program usually contains hundreds or thousands of binary rules, which are processed in sequence to lead to the desired action. The difficulty is that programs based on a binary thinking try to model the world in binary ways, but the world is not like this. It is fuzzy.

Another problem is that it has become apparent that the human brain does not operate in the crisp, binary way. With fuzzy logic the rules can be more widely cast and we need fewer of them. We do not have to know the detailed equations. We do not have to force everything to be either black or white, for fuzzy logic can easily cope with shades of grey. Combining fuzzy logic with neural networks (Chapter 7) promises to provide a closer approach than AI to designing the ultimate 'smart machine'.

Keeping up?

4 Write a definition of the crisp set N of even numbers from 32 up to but not including 44.

5 What is the membership degree of every member of a crisp set?

6 In what way does a fuzzy set differ from a crisp set?

7 What is a membership function?

8 Draw a diagram of a trapezoidal fuzzy set and include some of its members.
9 Give three examples of linguistic variables.

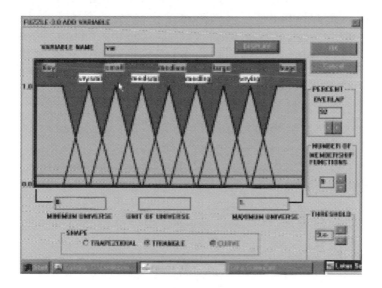

Figure 6.6 *As an example of software for building fuzzy logic systems, this screen from FUZZLE shows a range of nine linguistic variables from 'tiny' to 'huge'.*

Probability and membership degree

A diagram of a fuzzy set looks very similar to the probability distributions so often used in conventional statistical analysis. The resemblance is strongest if the membership function is defined by exponential curves. Then we obtain something looking remarkably like the standard bell-shaped Gaussian distribution. However, probability and membership index are very different.

Probability deals with random selections from a population of varying individuals. It tells us the probability of selecting an individual of given size, weight or other characteristic. To measure the parameters of a population, we first measure a large number of individuals exactly, group them into categories or classes (of size, weight and the like) and count how many individuals fall into each category. So the population distribution (Fig. 6.7) summarises the data gathered from a fair sample of a population of individuals. The distribution is made up of a number (often a very large number) of crisp sets. Either an individual falls into a given class (set) or it does not. It can not belong to two classes (sets).

For example, a probability distribution may plot the lengths of tails of mice. Tails are measured and individuals grouped into crisp sets (tail length 40-41 mm, tail length 41-42 mm, tail length 42-43 mm ...). A histogram of the results is plotted, the height of each column being

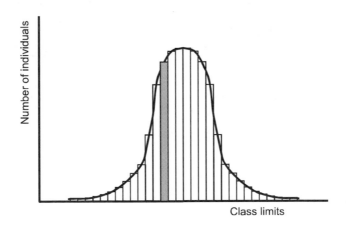

Figure 6.7 *A statistical probability distribution is based on the numbers of individuals in a sample that belong to each of many crisp sets or classes. The shaded set consists of individuals (perhaps several hundreds or thousands, if the sample is large) whose size (example, tail length) is less than average. Fewer individuals belong to this set than belong to the largest sets at the centre of the distribution.*

proportional to the numbers of mice in each set. A curve may be drawn, smoothing off the irregular tops of the columns of the histogram to give an overall view of the population distribution. The population distribution curve in Fig. 6.7 represents the whole sample of mice and, if the sample is sufficiently large may be taken with a predictable degree of precision to represent the whole world population of mice. Note that the 'tails' (!) of the distribution do not touch the x-axis. There is always the possibility that mice may yet be found with tails much shorter or longer than those found in the sample.

Fuzzy sets define categories that may or may not have many (or even any) members. They are defined so that we are able to say what we want to happen if an individual of a particular size happens to be found. We do not collect masses of data before defining the sets. Moreover, fuzzy membership functions allow individuals to belong to two or more categories at the same time, and often to a different extent to each.

For example, in Fig. 6.8 there are two Gaussian *fuzzy sets,* overlapping so that mice with tails slightly longer than medium may also belong to the set of mice with slightly longer tails. A mouse with tail length x

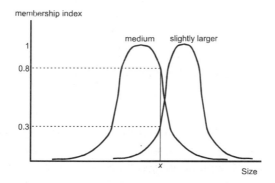

Figure 6.8 *Two fuzzy sets are overlapping, allowing an individual (represented by the vertical line) to belong to both sets.*

(maybe $x = 93$ mm) belongs with a membership degree of 0.8 to the fuzzy set of medium-tailed mice. It also belongs to the fuzzy set of slightly longer tailed mice with a membership degree of 0.3. Note that with Gaussian fuzzy sets, the 'tails' extend to infinity in either direction. This means that all mice belong to all sets. However, the membership functions of those mice on the extremes of each set are exceedingly small and can be ignored.

In Fig. 6.7, the height of the probability curve depends on the number of individuals in the largest class. In Fig. 6.8, the height of a membership function is always 1, the membership degree of a full member of the set.

From input to output

As a simple and hypothetical example, we follow the initial stages of designing a fuzzy control system for a clothes washing machine. Several such machines are on the market but this account is not related to any of them. The washing machine has a sensor that measures the dirtiness of the load. This is done by sensing the opacity of the washing water a few minutes after the wash cycle has begun. The sensor reading is from 0 to 100 on an arbitrary scale.

The input to the system is a reading from the optical system. Even though the scale may be arbitrary (it need not be in some systems), the input has a precise value. It is crisp. The next step before the analysis is to *fuzzify* it. Fuzzification consists of assigning the input value to one or more fuzzy input sets. The range 0 to 100 is divided into five fuzzy sets, named by linguistic variables (Fig. 6.9). If the sensor reading is 36, for example, it is a member of two sets, 'slightly soiled' and 'dirty' with a different membership degree for each.

The output variable is agitation. The machine can deliver a variable amount of agitation ranging on an arbitrary scale from 0 to 5 (Fig. 6.10). Agitation is divided into seven fuzzy sets, from 'very light' to

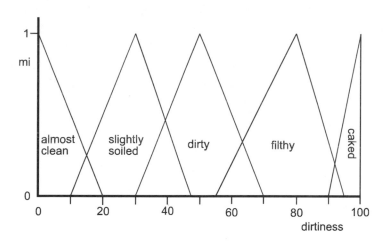

Figure 6.9 *The five fuzzy input sets of the washing machine example cover a range of dirtiness from 0 to 100.*

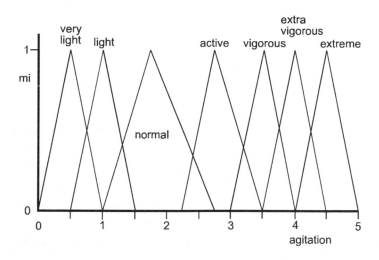

Figure 6.10 *The seven fuzzy output sets of the washing machine example cover a range of agitation strengths from 0 to 5.*

'extreme'. The higher levels of agitation must not be used unless the load is very dirty, for this would cause undue wear of the fabrics.

We can obtain a picture of the relationship between dirtiness and agitation by plotting a diagram like Fig. 6.11. This illustrates the five *fuzzy rules* that govern the controller:

1 IF dirtiness is 'almost clean', THEN agitation is 'very light'.
2 IF dirtiness is 'slightly soiled', THEN agitation is 'light'.
3 IF dirtiness is 'dirty', THEN agitation is 'normal'.
4 IF dirtiness is 'filthy', THEN agitation is 'vigorous'.
5 If dirtiness is 'caked', THEN agitation is 'extreme'.

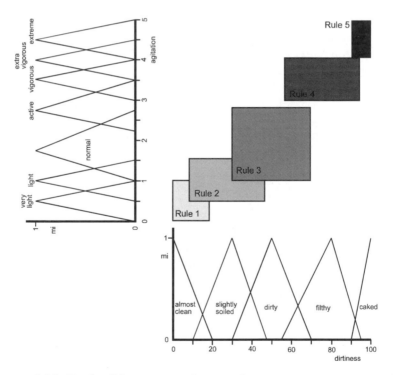

Figure 6.11 *Each of the rectangular patches represents a range of crisp input values and the corresponding range of crisp output values as defined by the five fuzzy rules. In very general terms, the dirtier the washload, the stronger the agitation.*

All rules are of the IF … THEN … type. There are five rules, one for each level of the input variable. Note that, as the system stands at present, there is no way in which agitation can be 'active' or 'extra vigorous'. Each rectangle in Fig. 6.11 indicates that, if the input is within the range specified by the width of the rectangle, the output must be within the range specified by its height. These rectangles are 'patches' that cover the relationship between the input and output of the system. The alignment of the patches shows that, as dirtiness increases, so must agitation increase. However, there is no attempt to impose a mathematical relationship. The relationship is just a 'common sense' one. The patches *might* be covering a curve showing an exact relationship, as indicated in Fig. 6.12. However, we do not have to know what form the relationship (if any) takes. The curve could be a straight line, an exponential curve, or a sinusoid. It does not matter what it is.

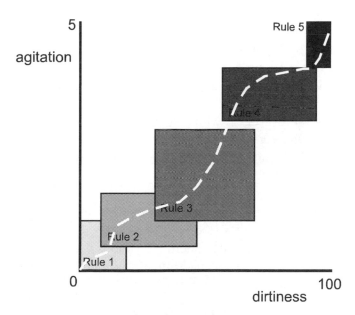

Figure 6.12 *The five patches of Fig. 6.11 may cover a more precisely defined relationship (dashed white line) between dirtiness and agitations, but we do not need to know if it exists or what the equation of it is like.*

Given the crisp input to the system, we convert this to fuzzy input. This is known as *fuzzification*. The first stage of fuzzification is to 'fire all rules'. In a typical AI program, the rules are processed in sequence. In a fuzzy program, they are all fired at once. Actually, this is not possible using a digital computer for its programming is sequential. However, we can arrange for the computer to produce the same effect as simultaneous firing.

Using Fig. 6.9 and the list of rules given above, the firing process with a dirtiness input of 36 is:

1 IF dirtiness is 'almost clean', THEN agitation is 'very light'.
2 IF dirtiness is 'slightly soiled', THEN agitation is 'light'. FIRED
3 IF dirtiness is 'dirty', THEN agitation is 'normal'. FIRED
4 IF dirtiness is 'filthy', THEN agitation is 'vigorous'.
5 If dirtiness is 'caked', THEN agitation is 'extreme'.

Two of the rules (2 and 3) are fired because the value 36 is a member of two of the fuzzy input sets. It is not a member of any other sets, so rules 1, 4 and 5 are not fired. Fig. 6.13 shows the extent to which Rules 2 and 3 are fired. For Rule 2, a vertical line through dirtiness 36 cuts the curve at membership degree 0.66. In other words, dirtiness 36 is a member of the set for Rule 2, with a degree of 0.66. At the same time, it is a member of the set for Rule 3, with a degree of 0.29. It is more of a member of Rule 2 than of Rule 3 but nevertheless both rules must be applied.

Having fuzzified the input and made the appropriate inference about the applicability of the rules, the next step is to apply the rules in proportion to their applicability. There are various ways in which this is done. Some of the ways give very similar results and some are better suited to certain types of fuzzy system.

In Fig. 6.14, we see the results of firing the rules. The sets for five of the agitation levels are not fired, so are ignored. Rule 2 had an applicability of 0.66 and so its output set (light agitation) is truncated at the 0.66 level. Similarly, the output set for Rule 3 (normal agitation)

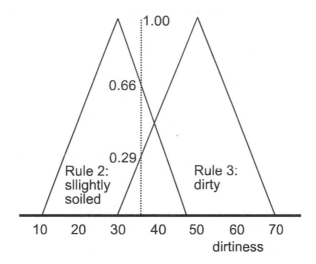

Figure 6.13 *A crisp dirtiness input of 36 fires both Rule 2 and Rule 3, but Rule 2 is more applicable than Rule 3.*

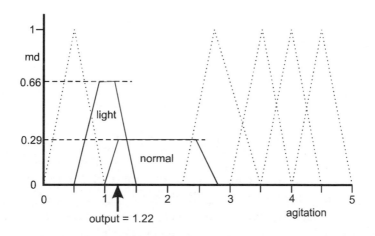

Figure 6.14 *The weighted crisp output of 1.22 is the result of combining the fuzzy outputs in proportion to the applicability of the rules.*

Simple rules

'Comfortable temperature' is a function of actual temperature AND wind speed. A heating control system based of crisp sets might have a rule that says:

IF the temperature is below 20°C AND the wind speed is more than 20 km/h, THEN turn on the heater.

Unfortunately, this leaves the heater off if the temperature is 21°C and the wind speed is 60 km/h. It also leaves the heater off if the temperature is 19°C and the wind speed is 5 km/h. It needs a series of dozens or even hundreds of rules to cover all possible cases in which the heater should be switched on, or partly on. Fuzzy logic can cover all contingencies with a single, simple rule:

IF it is cool AND windy, THEN turn on the heater.

is truncated at level 0.29. The fuzzy output of the system thus consists of the sets for light and normal agitation, with each being represented in the same proportion as the corresponding input sets.

It now remains to obtain a crisp output from Fig. 6.14. This is called *defuzzification*. Again, there are several techniques in use. One of these is known as the 'centre of area' or 'centre of gravity' technique. If you imagine the area of the pair of truncated triangles of Fig. 6.14, with the overlapping area cut out, this is the area of which the centre of gravity is to be found. This can be calculated by a relatively simple formula, but it is easy to gauge by eye that the centre of gravity lies on the vertical line at agitation 1.2. Therefore, 1.2 is the crisp output when the crisp input is dirtiness 36.

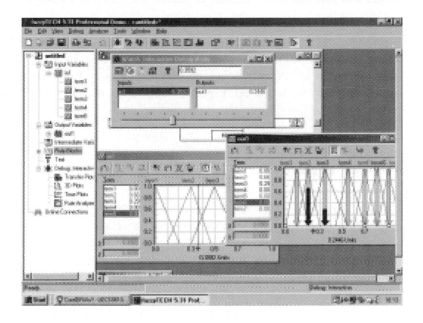

Figure 6.15 *Running the washing machine data on fuzzyTECH software produces a display like this. The IN1 window (lower centre) shows the fuzzy input sets. The Watch window (top centre) allows the user to vary the crisp inputs over their whole ranges. The corresponding crisp outputs are instantly displayed. Here a crisp input of 0.3592 on a scale 0 to 1 (equivalent to 35.92 on a scale 0 to 100) gives a crisp output of 0.2446 (equivalent to 1.22 on a scale of 0 to 5). The OUT1 window shows the weighting of the fuzzy outputs for Rules 2 and 3 (compare with Fig. 6. 14).*

Summing up, the stages in fuzzy control are:

- Crisp input.
- Calculate fuzzy input (fuzzification)
- Fuzzy inference.
- Fuzzy output.
- Calculate crisp output (defuzzification).

Calculating crisp output can be somewhat complicated, but fortunately, there is software available to perform this task. It may use the centre of gravity technique or one of the others and it is usually possible to select which is used. Fig. 6.15 shows the result of working the washing machine example given above.

Another input

For comprehensive control, a washing machine needs several inputs. We will add a second input to the controller. This is the *oiliness* of the dirt in the load, and may be estimated by measuring the rate at which the opacity of the washing water increases at the beginning of the wash. The oilier the load, the more slowly is the dirt put into suspension. Oiliness can be specified on a scale from 0 to 1, as in Fig. 6.16. The distribution of the sets is very asymmetrical, because it is thought that the exact amount of oil present is more important as the amount of oil increases. So the sets are narrower (more precisely defined) as the oiliness increases.

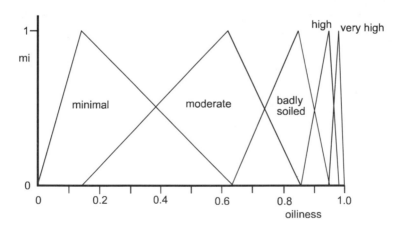

Figure 6.16 *Fuzzy input sets for the washing machine discriminate more between degrees of oiliness at the top end of the scale.*

Given two inputs, each with five levels, they can be combined into a set of rules of the form IF A AND B THEN C. This gives a total of 25 rules, which are summarised in this table:

	minimal	moderate	badly soiled	high	very high
almost clean	very light	very light	light	normal	active
slightly soiled	light	light	normal	active	vigorous
dirty	normal	normal	active	vigorous	extra vigorous
filthy	active	vigorous	extra vigorous	extra vigorous	extreme
caked	vigorous	extra vigorous	extreme	extreme	extreme

The columns represent the amounts of oiliness and the rows the levels of dirtiness. The body of the table lists the agitation levels for each combination of oiliness and dirtiness. For example, the top left cell represents:

Rule 1: IF oiliness is 'minimal' AND dirtiness is 'almost clean', THEN agitation is 'very light'.

The operation of the rules can be visualised by plotting a three-dimensional graph (Fig. 6.17) in which the plane shows the relationship between the two inputs and the output. In general terms, the plane slopes up (higher agitation) with increase of dirtiness and increase of oiliness. The effect of increasing oiliness is the greater. For particular pairs of crisp inputs, co-ordinates plotted on the graph show the corresponding crisp output. In the figure, arrows on the co-ordinates show that if dirtiness is 30 (slightly soiled) and oiliness is 0.8 (moderate and badly soiled), then agitation is 1.6 (normal). This is the result of firing the two rules which have blacked-in cells in the table above.

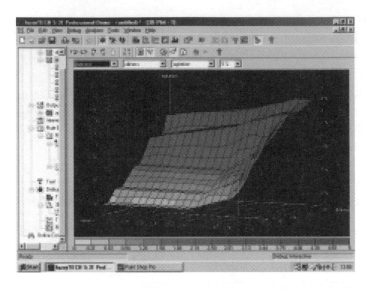

Figure 6.17 *An output plot in three dimensions by fuzzyTECH™ software shows the agitation level for all input combinations of dirtiness and oiliness.*

Here we leave the description of the hypothetical washing machine. It would be better with more inputs (such as thickness of fabric, efficacy of washing powder). It could also be designed to have more than a single output, such as length of wash time, and temperature of water. However all these refinements can be provided by simple extensions of the principles of fuzzy controllers as outlined above.

Approximating to conventional control

While extolling the virtues of fuzzy logic, it must not be forgotten that in many circumstances the conventional control systems described in earlier chapter may be perfectly satisfactory. If so, there is no need to turn to fuzzy systems. However, conventional systems depend heavily on the designer having a reliable model to base the system on. The mathematical equations underpinning the system need to be known.

Unfortunately, it often happens that the model is unknown or only imperfectly known. There was an instance of this in description of the power station steam generator (Chapter 2). The interactions between the various sub-systems are too complicated for the operator to be able to predict certain aspects of the behaviour of the system. Similar problems arise in controlling a blast furnace, and these have been solved using fuzzy logic. It is possible to design fuzzy systems that perform in a similar way to typical PD, PI and PID controllers. For instance, to build a fuzzy controller with an action similar to a PD system we provide it with two inputs:

* Error – the *difference* between the output and the set point.
* Rate of change of error – the *differential* of the error.

In the fuzzy system, the inputs will be crisp, but are converted to fuzzy inputs. Thus, error inputs might be grouped into five fuzzy sets:

large negative, small negative, nil, small positive, large positive.

Rate of change of error will be similarly grouped. There will also be fuzzy outputs sets describing how the output is to be varied:

large decrease, small decrease, no change, small increase, large increase.

Making the rules

As in the washing machine example above, rules can be statements of common sense. However, the system will probably give better results if expert advice is sought when preparing the rule block. In other words, a fuzzy system can be built up in much the same way as an expert system. It is the way that the system operates that is different.

Another approach is to watch and record the action of an experienced person operating the system manually. After a few trials, a number of fuzzy rules are proposed. These are built into a system that can then be tuned by running it on software of the kind used in our examples. The boundaries of the fuzzy sets can be adjusted, the rules can be amended and some additional rules added to the system. All of this is relatively easy to achieve with a fuzzy controller.

A number of other techniques are available for building and tuning fuzzy systems. These include:

- **Self-organising controller:** The controller evaluates its own performance and modifies its parameters to improve its performance.

- **Adaptive controller:** The controller evaluates its performance and uses this information to modify its own design.

- **Learning controller:** The fuzzy system is linked with another system that acts to improve its performance. This other system may be another fuzzy controller, a *neural network* (Chapter 7) or a *genetic/evolutionary algorithm*.

Genetic and evolutionary algorithms mimic biological processes. By methods involving random selection, the controller is changed (or mutated) and the performance is assessed. Favourable mutations are incorporated into future versions of the controller, while unfavourable mutations are discarded. The genetic algorithms include an operation analogous to crossing over, in which features of two related controllers are recombined with the hope of producing an even more effective controller. The process is repeated many times (that is for several generations) until the controller reaches a predetermined degree of fitness for its intended purpose.

Keeping up?

10 List the main stages between the input to and the output from a fuzzy controller.

11 Write a list of linguistic variables that might cover the input to the fuzzy controller of a dishwasher.

12 What is meant by fuzzification?

13 What is fuzzy inference?

14 In what order are the rules fired in a fuzzy controller?

15 What is meant by defuzzification?

Fuzzy logic in action

There are so many successful applications of fuzzy logic to control systems that it is possible to describe only a small proportion of them. Among the more interesting are:

Washing machines with fuzzy control have been marketed by several companies including Daewoo (Korea), Goldstar (Korea), Hitachi (Japan), Matsushita (Japan), Samsung (Korea), Sanyo (Japan), and Sharp (Japan). These machines adjust the washing strategy according to dirtiness, fabric type, load size and water level. The outputs of the controller include the quantity of water used, the water flow speed, the washing time, the rinsing time and the spin time. Some models use a neural network to tune the processing rules to the user's preferences. The machines are claimed to wash more quickly, with significant savings in the amount of electrical power used.

Air conditioners with fuzzy control have greater stability both as coolers and as heaters. There is less fluctuation in room temperature than with conventional PID control. Mitsubishi (Japan) claim that their air conditioner uses less power than a conventional machine, saving 24% of energy costs when cooling and 17% when heating.

Automobile engine management by fuzzy logic controls both fuel injection and ignition. In the Nissan (Japan) system control is based on throttle setting, oxygen content, cooling water temperature, engine RPM, fuel volume, crank angle, knocking, and manifold pressure.

Fuzzy chips

Most fuzzy controllers are operated on a typical PC or make use of a programmed microcontroller chip. This has the advantage that there is plenty of relatively inexpensive hardware to choose from when implementing a system. Software for designing and debugging fuzzy logic controllers is generally written to run under Windows on a typical PC. Two examples demonstrated in this chapter are *fuzzyTECH* and *Fuzzle.*

As has been explained, conventional computers employ serial processing, while fuzzy logic required parallel processing ('Fire all rules'). This makes conventional computers inefficient for running fuzzy logic, and even more so if the fuzzy controller is associated with a neural network. There is thus a need for hardware specially designed for fuzzy processing. Various chips have been produced, including some based on analogue circuitry. The first successful digital fuzzy processor was the Togai FC110. A newer chip is the Adaptive Logic™ AL220. It has four analogue inputs with 8-bit resolution and four analogue outputs also with 8-bit resolution. The chip includes units for fuzzification and defuzzification and can process 500 000 rules per second.

Following the development of an increasing range of fuzzy chips, the next obvious development has been the introduction of PLCs based on fuzzy logic. The fuzzyPLC™ is programmed on-line by using the fuzzyTECH software (produced by Inform Software) that was used in this book for Figs. 6.15 and 6.17.

Microwave ovens with fuzzy controllers are made by the Japanese firms Hitachi, Sanyo, Sharp and Toshiba. Sensors measure the temperature and humidity within the oven and the controller determines power levels and cooking times accordingly.

Television sets made by Hitachi (Japan), Samsung (Korea) and Sony (Japan) adjust the screen colour balance and texture for each frame. Sound volume is stabilised according to the acoustics of the viewer's room. In some TV sets, the fuzzy controller monitors the relative brightness, the contrast and colour balance of each frame. The parameters of the TV tube are then adjusted frame by frame to obtain optimum appearance.

Camcorders suffer from camera shake when held in the hand. Matsushita (Japan) have a camcorder that eliminates this problem to a high degree. Each frame is compared with the previous one. If only parts of the image have moved, this is just normal movement of objects within the field of view. Such a frame is left untreated. However, if most objects have (apparently) moved by about the same amount and in the same direction, this is evidence of shake. The current image is displaced to match the previous one. Automatic focusing of the Sanyo (Japan) 8mm fuzzy video camcorder FVC-880 is based on only 9 fuzzy rules. The image produced by the lens is divided into 6 regions. The central region has the most influence on the result and the border regions least. The relative contrast between regions provides the information needed to focus the lens.

Lifts with fuzzy programming cut average waiting times to 50%. Input to the controller includes an estimate of the sizes of the groups waiting for the lift, the time of day, and the distance the lift has to travel.

Traffic control in the Japanese city of Nagoya (population 2 million) was installed by Matsushita and is an early example of a successful fuzzy logic system.

Subway control at Sendai (Japan) was one of the first (1987) examples of a large engineering system with fuzzy predictive controls. The system controls the acceleration, running speeds and stopping of the trains, with the aim of improving safety, accurate stopping at the platforms, reducing running times, reducing energy consumption, and providing maximum comfort for the travellers. The system controls not only the locomotives but also other aspects of a railway system such as route control, public address systems, and ticket sales.

Test yourself

1 Fuzzy logic was invented by:

    A Ebrahim Mamdani.
    B John Venn.
    C George Boole
    D Lotfi Zadeh.

2 Boolean logic is:

    A fuzzy.
    B multi-valued.
    C true.
    D binary.

3 In crisp logic, the rules are applied:

    A in parallel.
    B in series.
    C in any order.
    D in order of importance.

4 In a fuzzy set, the members have a membership degree:

    A of 0.
    B less than 1.
    C between 0 and 1.
    D that is fuzzy.

5 In a triangular fuzzy set, the number of members with a membership degree of 1 is:

    A 1.
    B about half of them.
    C none.
    D all of them.

6 The inputs to a fuzzy controller are:

    A fuzzy.
    B crisp.
    C binary.
    D linguistic variables.

7 Obtaining a crisp output from the fuzzy output sets is called:

    A fuzzy inference.
    B firing the rules.
    C defuzzification.
    D the centre of area technique.

8 A fuzzy controller which has fuzzy inputs for error and the time integral of error, can be made equivalent to a conventional:

    A bang-bang controller.
    B PID controller.
    C PI controller.
    D PD controller.

9 An adaptive controller:

    A modifies its own parameters.
    B is trained by a neural network.
    C is based on a genetic algorithm.
    D modifies its own design.

10 A fuzzy controller does not need:

    A an equation to relate output to input.
    B crisp input.
    C rules.
    D fuzzy output.

7
Artificial neural networks

As systems become increasingly complex, the routines for controlling them become increasingly elaborate. In this context we are referring to systems of the complexity of power stations, blast furnaces, landing lunar modules, countywide traffic systems, and humanoid robots. Much research has gone into finding the best ways of controlling such systems. One way is to put a full-time human in charge, for there are few computers or electronic circuits that can out-perform the human brain as a general-purpose controller. However, complex tasks may require exceptional knowledge or skills, perhaps both. Persons with the necessary combination of knowledge and skills for a given control task may be in short supply.

Some tasks require rapid intake of data and rapid response – perhaps faster than the human brain can provide. In the medical field, the shortage of skilled and experienced physicians has led researchers to develop computer programs that question a patient and diagnose the disease. The control of spacecraft trajectories in real time gives further scope for employing computers, as do the control of industrial plant, and the preparing of weather forecasts. Allied applications are the recognition of patterns, from fingerprints to faces, from financial market trends to handwriting.

Artificial intelligence (AI) researchers have developed the *expert system,* in which the knowledge and skills of an expert (say, a consultant opthamologist or a top financial adviser) are embodied in a computer program. The user is questioned and feeds in data. The program produces the same answers that the expert would give when provided with the same data. Such programs are generally based on formal logic. In programming terms, it is a mass of decisions of the IF... THEN type, encountered sequentially. On a much smaller scale, we come upon this procedure when we pay the telephone bill by telephone. In response to a sequence of pre-recorded questions, we key in numbers and are gradually routed to that section of the accounts department that is concerned with accepting payment. It is a matter of narrowing down the possibilities by a sequence of bivalued or multivalued choices. In the same way, a medical expert system can narrow down the possible diseases to one particular disease or perhaps, where there is uncertainty, to a group of diseases.

The mention of uncertainty reminds us that in real life a situation is seldom as clear-cut as we would like it to be. Real life systems do not necessarily match the binary nature of most computer systems. This is why the advances in fuzzy logic (Chapter 6) are so important and rewarding. Unfortunately, even fuzzy logic can not provide all the answers.

Since the human brain is still the best for complex control operations (except for certain specialist applications and, even then, it is a human brain that writes and tests the programs) we should look at how it works. The first step is to look at its structure. The brain is built up from nerve cells, or *neurons,* each of which has a structure similar to Fig. 7.1. Each neuron consists of a single cell. The rounded cell body is microscopic and is complete with a nucleus. Radiating from this are two sets of branching structures, the dendrites and the axon. The dendrites are relatively short but, in certain types of cell such as those conveying messages from the spinal cord, the axon may be a metre or more long. The axon ends in a tree of branches each ending in a swollen synaptic button.

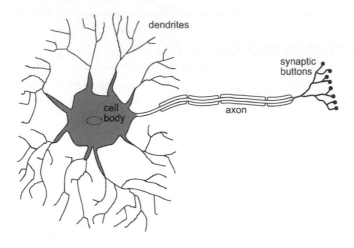

Figure 7.1 *A neuron transmits nerve signals in one direction only: from the dendrites, through the cell body and axon, to the synaptic buttons.*

The dendrites are in close contact with the synaptic buttons of other neurons and receive nerve messages from them. The messages take the form of changes in the cell membranes, which cause changes of potential across the membrane. They cause electrical signals to pass along the surface of the dendrites, the cell, and the axon. The messages pass inward along the dendrites to the cell body and then out along the axon to the synaptic buttons. These are located close to the dendrites of other neurons. In this way, the neurons build up a neural network (Fig. 7.3), mainly in the brain, but also in other parts of the nervous system (such as the retina of the eye). The human brain has about 100 000 000 000 neurons and each of these can connect to as many as 10 000 other neurons. This gives a total of about 10^{16} possible synapses in the brain. Learning and memory depend upon the actual connections that are made at the synapses. With 10^{16} possibilities it can be seen that the brain potentially has a very large capacity for learning, and making responses based on that learning. In the world of control systems the brain of a human controller is made aware (perhaps by a display like that on p. 52) of the conditions obtaining in different parts of the system and operates the control levers and buttons accordingly.

Recognising patterns

The brain has remarkable faculties:

- It can recognise patterns – faces, signatures, voices, music, and words.
- It can recognise incomplete patterns – a snatch of song, a face half in shadow, a sentence when the bottom half of each word is covered.
- It can recognise distorted patterns – a voice speaking on the telephone, an elongated shadow, the same letter printed in different fonts, and almost illegible handwriting.
- It can recognise patterns against a background of noise – a conversation at a busy traffic centre, a face lit by disco lighting.

Artificial neural networks too can recognise patterns, even when they are corrupted for some of the reasons given above. This is because ANNs need not rely on precise, formal rules.

The brain is so complex that we are only just beginning to understand its simplest levels of operation. This is why ANNs and neurocomputers can not be said to be based on our knowledge of brain structure and function.

In this chapter, the term *neural network* is restricted to the natural neural network found in the nervous systems of humans and other animals. The so-called neural networks developed in hardware or in computer systems are referred to as *artificial neural networks*, shortened for convenience to *ANN*. The distinction is important because, although ANNs may appear to have some of the features of neural networks, the similarities are superficial at best. No one can claim to have built an ANN that works in the same way as a human brain, or has a performance equal to a human brain. It is possible that such a device will never be built. However, in spite of this, it is sometimes helpful to draw comparisons, as in the next paragraph.

One is tempted to compare the brain with a computer hard disk capable of storing 10^{16} bits. In other words, two million 5 Gbyte hard disks. To do this is misleading, for the brain does not store its data as a bit or bits in a neuron. The information in the brain is held in the *connections* between the neurons, not in the neurons themselves. Learning consists of building up the connections. Note that in this chapter we use the word *neuron* for both natural and artificial neurons, as the kind intended is usually clear from the context.

Because a given item of data is not stored in a particular neuron, neurons can be removed from the brain without loss of memory. The connection patterns permeate the whole brain. The situation is similar to that of a 3D hologram, in which the image is stored throughout the hologram, not in a particular area of it as in a conventional photograph. We can remove a portion from the hologram yet still see the whole image, though with a more restricted viewing angle.

Keeping up?

1 What are the parts of a human neuron, and in what direction do the nerve messages pass?

2 What do we call the junction between a synaptic button of one neuron and the dendrites of another neuron?

3 Approximately how many neurons are there in a human brain?

4 In what way does the brain store information?

Learning

We have said that learning involves the building up of connections between neurons. It is not just a matter of which connections are made but also the strength of the connections. The connections may be either excitatory or inhibitory. When the connection is excitatory, a message passing in through a dendrite makes a response more likely to occur. An inhibitory connection makes the response less likely. Further, the strength of the signals varies with the synapse, so some signals are strongly excitatory while others are more weakly so. At any given instant, a neuron is receiving signals, some excitatory, some inhibitory, some strong, some weak, from the neurons to which it is connected. The cell body totals these signals and, if the signal exceeds a threshold value, it passes a signal out along its axon to the cells in contact with its synaptic buttons. Thus learning involves making some connections stronger and some weaker. We say that the connections are *weighted*, so that messages arriving from some sources have more influence than others on whether or not the neuron passes a signal on.

Electronic counterpart

It is possible to build a circuit that has many of the features of a neuron. In Fig. 7.2 the cell body is represented by an operational amplifier, connected as an adding circuit. If a number of different voltages are applied as inputs, the output of the op amp is inversely proportional to their sum. Each input first passes through a resistor. The resistors weight the inputs so that, if the inputs are actually logic levels their effect on output depends on the value of the resistor. In a more realistic model, the resistors would be variable, so that the weightings could be changed as the neuron learns. A resistor might be a potentiometer driven by a small electric motor. This would be used to vary the resistor continually as the network is being trained.

The output from the op amp is passed to an inverting buffer with a

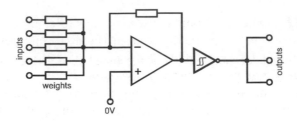

Figure 7.2 *An operational amplifier wired to sum weighted inputs has some, but not all, of the characteristics of a neuron.*

Schmitt trigger output. The output from the op amp normally swings negative and the output of the buffer goes to logic high whenever the voltage falls below a certain threshold. This models the action of the cell body in generating an output signal when the total input signals exceed a given threshold. The output from the buffer is then fed to a number of other neurons.

The circuit may help you to understand what a neuron does, but a number of refinements would be needed to make this a satisfactory model. In one respect, it is completely unlike a neuron. The signals produced by a neuron are electrical, but they are not voltage levels. The neuron generates *pulses,* which is something that this circuit does not do. The rate of pulse production, the *rate of firing,* is a measure of the strength of the signal.

Building a hardware network from electronic neuron models like Fig. 7.2 is a daunting task. To realise even a simple learning network requires large numbers of neurons. In the days when the operational amplifier had to be constructed from thermionic valves, the circuits were very limited in ability and reliability. With the arrival of solid-state devices, they become more practicable. Hardware ANNs are now made as custom-designed integrated circuits, using very large scale integration (VLSI, the same level of complexity as the Pentium and other microprocessors). Although experimental ANNs had been built before the advent of VLSI, it was not until VLSI arrived that real progress was made.

A true *neurocomputer* is a computer with architecture based on ANNs. Such a computer differs markedly from the familiar digital computer. A digital computer is *programmed* to accept data from the outside world and *programmed* to store it *symbolically*. Its *program* then processes the symbolic data. Processing is *serial*. A neurocomputer accepts data from the outside world and sends it directly to the inputs of its neural net. It then processes it in parallel, applying the weightings that it has been *taught* or *trained* to use.

Each neuron in the ANN of a neurocomputer has its own *local memory*. This stores a quantity that is available only for computations associated with that particular neuron. It can be summed with the signals arriving from the dendrites. In other words, the behaviour of the neuron is influenced to a certain extent by its previous experiences. Conversely, the value in the local memory can be updated when the neuron produces output.

The program of a digital computer must necessarily be based on a known set of rules and on known procedures. For many cases, for example the control of a power station, it may be that not all the rules are in fact known. In any case the algorithms for controlling the station are highly complex and interact to further increase the complexity. By contrast, a neurocomputer does not need to know the rules. It can link a required output to a given input without having to know the rules (if any are definable!) that connect them.

Mathematical models

As an alternative to the electronic model neuron, we can simulate a neuron mathematically. We can write an equation showing the inputs and their weightings on one side and the output on the other. We can construct a virtual network (Fig. 7.3), feeding the output from a neuron to a number of other neurons. The model is complex and tedious to work on paper, but it can be converted into a program that will run on a digital computer. A simple example of this approach is *Easy NN* (p. 225).

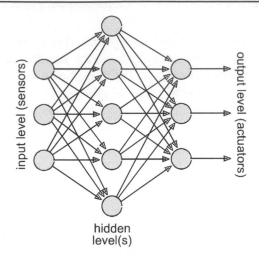

Figure 7.3 *An artificial neural network may take several different forms. A form often used in research is the feedforward network shown here. It has three or more levels of neuron. Each neuron sends signals to the next layer in the input-output direction.*

The following is an example of an ANN that might be used for controlling an industrial process. The example is a trivial one that could be equally well realised by using a few logic gates or a single logical statement in a BASIC program. However, an example that is sufficiently complex to do justice to the ANN technique would be difficult to understand without deep study. The example concerns a reaction vessel in which a mixture of substances is stirred and heated until the reaction has occurred and the mixture has been converted to a viscous compound. There are four sensors in the vessel, each of which gives a binary output:

- Cold: its output = 1 when the temperature is below 40°C.

- Hot: its output = 1 when the temperature is above 45°C.

- Viscosity: its output = 1 when the coefficient of viscosity exceeds 100 Pa s.

- Full: a level sensor with output = 1 when the tank is full to capacity.

There are four actuators in this system:

- Heater: a valve that opens (=1) to allow hot water to circulate in a jacket around the vessel.

- Cooler: a valve that opens to allow cold water to circulate in the jacket.

- Stirrer: a motor that drives a stirrer.

- Done: an indicator lamp that is to light when the process is finished.

Four inputs and four outputs are enough for an example. A real industrial plant would have many more, but the principle of generating an ANN is the same. This ANN will have three levels:

- Input.
- Hidden.
- Output.

It will be trained by supervision, that is, we will tell the ANN what output is expected for every possible combination of inputs. Note that there is to be no formal causal connection (logical or otherwise) between input and output. The ANN is developed by the software, which establishes the weightings needed on all the connections, so as to produce the correct output in the manner of a neural network.

The next thing to do is to decide what outputs are to be expected. As this is basically a logical problem, the data can be set out in a truth table as shown overleaf.

The first four columns are the inputs. The table runs through all possible combinations of these, except that the cold and hot sensors can not both give a 1 input at the same time, so this combination need not be included. For each of the 12 possible inputs, we enter the desired outputs in the last four columns. Here *we* are being the 'expert', training the system to give the appropriate responses. Only in row 3, for example, do we have the vessel full of viscous liquid at a

Row	Cold	Hot	Visc	Full	Heater	Cooler	Stirrer	Done
0	0	0	0	0	0	0	0	0
1	0	0	0	1	0	0	1	0
2	0	0	1	0	0	0	0	0
3	0	0	1	1	0	0	1	1
4	0	1	0	0	0	0	0	0
5	0	1	0	1	0	1	1	0
6	0	1	1	0	0	0	0	0
7	0	1	1	1	0	0	1	0
8	1	0	0	0	0	0	0	0
9	1	0	0	1	1	0	1	0
10	1	0	1	0	1	0	1	0
11	1	0	1	1	1	0	1	0

temperature between 40°C and 45°C. The 'done' lamp lights for only this combination of inputs because it would give false results to measure the viscosity when the liquid is too hot or too cold. We do not want the heater on while the vessel is filling, so the heater comes on only when the liquid is cold and the vessel is full. Similarly, the stirrer comes on only when the vessel is full.

A table with this combination of inputs and outputs is keyed into the neural network program. Then the program is told to generate the ANN. This is done almost instantaneously with such a simple system. The program is usually left free to put in as many neurons as necessary in the hidden level, and in this case it decided on six. The program calculates all the weightings and reports that teaching has been completed. The network may now be represented by a diagram similar to Fig. 7.4.

The next step for the experimenter is to test the network with various combinations of inputs. This is done in a special window, which lists the inputs and outputs and allows the inputs to be edited. It is found that the combination of outputs obtained always agrees with the truth table. Only when the 'viscous' and 'full' outputs are '1' does the 'done' lamp come on, while the stirrer continues to turn.

In the actual plant, it might be thought important to measure other things, such as pressure. It might be desirable to add other actuators, such as a valve to drain the viscous fluid from the vessel. Adding another variable to a hardware logic circuit or to a computer program could cause difficulties. Adding extra columns and lines to the truth table is easy, and a new network is generated in a fraction of a second.

Input to an ANN is not limited to binary values. As the following hypothetical example shows, we can use the analogue signals from the

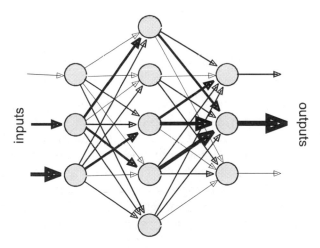

Figure 7.4 *The connections between neurons are initially of equal weight, as in Fig. 7.3 but, after the network has been trained, some are more strongly weighted than others. Then a given combination of inputs results in a learned combination of outputs.*

sensors after they have been converted to digital form. A clothes washing machine has a sensor that, soon after the wash begins, measures the opacity of the water, which is proportional to the amount of dirt in the clothes. This is expressed on an arbitrary scale from 2 to 20. There is a thermometer to measure the temperature of the water in degrees Celsius. Finally, the user keys in a value to indicate the average thickness of the fabrics, using a scale from 1 to 7. Thus, the ANN is presented with 3 numerical inputs. There is one output from the system, which is the required washing time in minutes. This can range from 4 to 30 minutes.

There are no mathematical equations that can calculate the interaction between dirt, temperature and thickness. Also there is no expert who could reliably estimate the required washing time. So the expert supervision of the previous example can not be arranged here. A practical approach is to discover suitable times empirically and to use these trials to train the network. In such a training session, several batches of clothing of differing dirtiness and thickness were washed at

Patterns in control systems

The concept of patterns may be extended to the input to a control system. For example, in an industrial plant, the pressure is high, the temperature is low, the quantity of product is slightly below average, and the viscosity of the product is excessively high. These four assessments (high, low, below average, excessively high) constitute a pattern that could be recognised by an ANN. It could process the data and initiate appropriate action.

Note that the input conditions are here expressed in fuzzy terms. ANNs do not need formal mathematical rules. Likewise, they can deal with fuzzy input quantities.

ANNs and fuzzy logic

Fuzzy logic systems can be designed by using ANNs. There is no need to specify precise rules and thresholds since ANNs can function without these. The network can simulate the fuzzy system, correcting its weightings until it performs as required.

Fuzzy systems may often use an ANN to continually adjust the boundaries of the fuzzy sets.

different temperatures. The operator ran the washing machine, checking the clothes every minute until they were judged to be satisfactorily clean. This was a subjective assessment but the best that could be managed in the circumstances. The data obtained were entered in the *Easy NN* program. Here are the first few lines of the data table, which holds details of 15 washes:

Wash no.	Dirt	Temp	Thick	Time
0	14	43	4	18
1	2	34	5	8
2	10	40	7	18

The details were entered into the computer and the program instructed to generate the ANN. It produced a net with three inputs and one output. The program was given free rein to generate as many neurons as it needed in the hidden level, and it set up six. The net was validated by entering some of the original wash input data, in response to which it produced the same output data. It was then tested by entering some new input data. In all cases except one, it calculated the correct wash times. The only exception was some extreme data at maximum temperature with thick fabric but, even then, the error was only 2 minutes.

Fifteen trials were found to be sufficient to train the network satisfactorily. If we had found an unacceptable degree of error after 15 trials, the error would probably have been reduced by including data from further trials.

Keeping up?

4 What are the two main types of connection between neurons?

5 Describe the features of a neurocomputer.

6 What is an ANN?

Network architecture

Several different kinds of network architecture are in use. The *feedforward network* illustrated in Figs. 7.3 and 7.4 is one of the simplest and most widely used. A feedforward network may have more than one hidden level and all levels may contain more than one neuron.

Many other architectures are possible and there is much research into the effectiveness of different types of network for use in different applications. Some of these other types of network have feedback to earlier levels, and there may also be connections between neurons on the same level. These different architectures are mainly experimental ones and do not necessarily bear any relationship to networks occurring in the brain.

Types of training

Training (or learning) may take three main forms:

- *Supervised training* consists of setting up a given combination of inputs and at the same time comparing the outputs obtained with the desired combination of outputs. This is the kind of training illustrated in the examples above. Training the network is automatic. Usually, small random weightings are assigned to the initial connections. The software runs the network once and the difference between actual and desired outputs is examined. As a result, the weightings are adjusted. The network is run again automatically with the new weightings and the process is repeated until the weightings stabilise with minimum error.

- *Reinforcement training* resembles the way in which a dog trained to perform a given task, such as acting as a guide to the blind, or searching baggage for drugs. The dog is rewarded with praise or food when it does what it is expected to do. As it learns more and more 'good' behaviour, it is rewarded and it adds a little more 'good' behaviour to its repertoire. Gradually, it learns the complete set of behaviour patterns for performing its allotted role. A network being trained by reinforcement is never told what outputs it is supposed to produce in response to a given set of inputs. Instead, it is given its reward in the form of a score. The value of the score depends on how well its outputs have matched up to the required pattern after a given number of runs. It uses this information to modify its weightings and it gradually acquires the desired behaviour. Reinforcement training is not so suited to control systems as is supervised training. The systems are usually complex and the network is unlikely to hit upon an appropriate response. It needs initial supervision, at least in the early stages of training.

Analogy – beware!

Remember that the analogy of learning in a dog is used only to illustrate the *idea* of reinforcement training. It does not imply that the *mechanism* is the same in an ANN as it is in a dog.

- *Self-organisation training* leaves the network on its own to discover what it is supposed to do and how to do it.

Control systems

The examples given earlier in this chapter show that an ANN can be trained to produce a given output as a result of receiving a given input. This ability is of obvious use in complex control systems. For example, a given set of input values (temperatures, light levels, proximity signals, gas pressures and the like) is detected by sensors that form the input level of the network. The data is processed by the network and the result is passed to actuators at the output level. These switch on or off to various extents to obtain the required control action. In such supervised systems, the output values used for training are generally set up as a result of expert advice. However, this can be advice based solely on experience. There is no need for it to be derived from expert knowledge of scientific laws or equations.

Sensor processing is one area in which ANNs are being used in industry. The input from a thermistor circuit needs little processing, but the input from a video camera needs a lot. Character recognition and the recognition of faces are important in many applications. In character recognition, the field of the camera is restricted so that it becomes filled with the character to be recognised. Then it is divided into grid of a hundred or more pixels. Each pixel may then be represented by 0 (paper) or 1 (ink). These provided the hundred or more inputs to an ANN. There are 26 outputs, each corresponding to a letter of the alphabet. By videoing different specimens of a given letter in turn, and telling the network which letter it is, the net can be taught to recognise a letter with a reasonable degree of reliability.

It is difficult to obtain detailed information of applications in working industrial control systems. A relatively short time has elapsed since suitable neurocomputers became available and much of the work with these has been experimental. No doubt many other applications are closely guarded trade secrets. The automobile autopilot devised by Shepanski and Macy in 1987 was a simulation in which an ANN learned from an expert driver how to negotiate itself along a two-lane highway on which other traffic was present. After the trials, it was considered that the ANN had the abilities of a reliable driver that would survive on a real road. No one seems to have put the ANN to a practical test. It is interesting to note that when different expert drivers were used to train the network, the network exhibited the distinctive driving style (aggressive, patient, hesitant) of the expert.

One well-known research topic serves to demonstrate the potentialities of control by ANNs. This is the broomstick balancing task (Fig. 7.5) originally devised by Bernard Widrow in 1962. It was made the subject of an ANN project in 1987. A broomstick is pivoted on a cart, which runs along a short straight length of track. The stick is pivoted at the far side of the cart (as viewed in the figure) so that it can hang vertically downward. The apparatus is viewed by a video camera and the image is analysed to find the present location of the cart, and the broomstick's angle from the vertical. From these measurements taken at regular time intervals, the velocity of the cart and the angular velocity of the broomstick can be calculated. The operator (whether human or artificial) is able to control the acceleration of the cart in either direction. The task is to start with the broomstick nearly vertically upward and then to balance the broomstick, keeping the cart

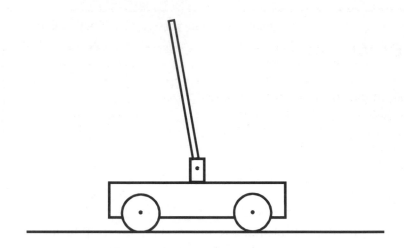

Figure 7.5 *The balancing broomstick experiment demonstrates the capabilities of ANNs.*

near to the centre of the track and to keep it balanced for a period of several minutes. A human operator could watch the cart and broomstick to decide how to vary the acceleration.

No human operator was able to succeed in this task. However, when a simulation of the apparatus was run at a tenth or less of real time, human operators were able to balance the broomstick. After being loaded with data from human slow-speed trials, the ANN could be taught the balancing technique. It was then able to balance it successfully in real time.

Keeping up?

7 Describe the main types of network training.

8 Describe an example of an ANN and how it is trained. What applications could there be for this network or one like it?

Breakdowns

A fault or breakdown in a single step of a conventional digital computer program is likely to lead to a complete breakdown. If a single step, such as an IF ...THEN statement, gives the wrong outcome, the subsequent processing is faulty and the result of the computation is seriously in error. It may even happen that the computer crashes.

A breakdown in a neural network is usually much less serious. If one of the neurons produces a faulty output, this is combined with the outputs from all the other neurons at the same level. The effect of the fault is diluted. The network does not stop working and the output is probably only slightly in error. Moreover, after a few more runs, the network may be able to compensate for the damaged neuron. The network is self-healing, in a similar way to a slightly damaged brain.

Summing up

A neural network, either natural or artificial, consists of connected neurons. Each neuron is a centre of information transfer. The extent to which the information is transferred from neuron to neuron depends upon what connections exist and the intensity of the connection.

A given pattern of input is matched by a pattern of output. ANNs can learn the connection patterns and so acquire a complex behaviour of use in control systems and other applications.

Test yourself

1 A program that asks the user questions and then gives legal advice is called a:

        A fuzzy system.
        B expert system.
        C artificial neural network.
        D neural network.

2 The human brain is made up of millions of:

        A axons.
        B cells.
        C corpuscles.
        D neurons.

3 Nerve messages pass directly from an axon to:

        A the cell body.
        B synaptic buttons.
        C dendrites.
        D the nucleus.

4 Information in a neural network is held in the:

        A nuclei.
        B cell bodies.
        C connections between neurons.
        D neurons.

5 A connection that makes a response more likely is said to be:

        A excitatory.
        B positive.
        C feedforward.
        D inhibitory.

6 A neurocomputer is distinguished by having:

        A memory.
        B a microprocessor.
        C parallel processing.
        D VLSI.

7 The input level in an ANN often consists of:

        A neurons.
        B data.
        C dendrites.
        D sensors.

8 Training an ANN alters its:

        A weightings.
        B input.
        C connections.
        D hidden level.

9 If an ANN is trained by 'rewarding' it for producing expected outputs, the training is described as:

        A feedforward.
        B feedback.
        C reinforcement.
        D self-organisation.

10 In contrast to a digital computer, a neurocomputer is:

        A programmed.
        B trained.
        C faster.
        D based on fuzzy logic.

8
Remote control

The essential point about a remote control system is that the operator is at some distance from the controlled device. In the case of an unmanned spacecraft cruising past Saturn, the controlled device, the spacecraft, is undoubtedly remote from its earthbound controller at Houston. In the case of someone relaxing in an armchair and switching channels on the TV, the controlled device is not as remote. However, it still qualifies as remote control. In the case of someone pressing a push-button by the front door and causing a radio-controlled bleeper to sound in the kitchen, this too just about qualifies. Yet, if we replace the radio link with a pair of wires from button to bleeper, we would probably not think of this as being remote control. It is hard to define exactly what we mean by remote control, but this chapter attempts to cover the more important aspects of the technology.

Typically, a remote control system has the following components, as depicted in Fig. 8.1:

- Operator.
- Transmitter.
- Receiver.
- Controlled device.

The transmitter is linked to the receiver through a communication channel, which as explained above may span a distance ranging from a few metres to a million kilometres or more.

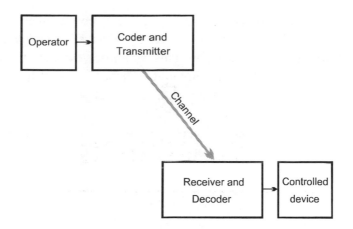

Figure 8.1 *In a typical remote control system, the transmitter sends a pulse code, which is received and decoded. Note that there is no electronic feedback in the simple system illustrated here. The controlled device is in view of the operator, so the control loop is completed by visual feedback.*

There are two points to notice about the system shown in Fig. 8.1:

- There is an operator. The purpose of a remote control system is to allow someone to exercise control from a distance (from the armchair, for example). In many of the control systems described in earlier chapters, the controller has been *automatic*. An operator is needed only from time-to-time to make occasional adjustments or to switch the system on and off. If a remote control system incorporates any automatic features, these are best provided by circuits close to the controlled device. For example, a spacecraft carries control circuits to automatically direct its solar panels toward the Sun and its antennae toward the Earth.

- There is no electronic feedback within the system. One reason for this is that remote control has a controlling function but not a regulatory function. If we want to keep the interior of an unmanned spacecraft at a constant level, we employ an *on-board* thermostat system. This simplifies the remote control system and reduces the amount of data to be sent back to the controller. Indeed, when the spacecraft is exploring other planets the signals take several hours to arrive. The length of time taken for radio signals to be sent from the craft to Earth and back would prevent any reasonable regulation to be attempted.

In most remote control systems, the feedback is purely visual. We watch the TV screen to see what channel has been selected and a graphic display on the screen tells us the present volume level. An enthusiast controlling a flying model aeroplane watches as it performs its aerobatics overhead. Visual feedback may sometimes use remote video cameras, as in devices handling hazardous waste, or in remotely operated vessels (see box). Video links may also be used where a system is widely spread across the country, as in the Merlin radio astronomy network (see boxes, pp. 190-1).

More recently, it has become unnecessary to use feedback from the actuator back to the main control system. Microcontrollers have become so inexpensive and with so many facilities, that such control takes place in or close to the actuator itself. This *distributed processing* was described in Chapter 5 in connection with the 'fly by wire' system used in the Airbus. Feedback is local and does not take place over the remote control channel.

Transmitters

There are four main ways in which commands may be transmitted:

Ultra-sound: The transmitter consists of an oscillator running at a frequency too high to be detected by the human ear. The ear can hear sounds up to 20 kHz and commonly an ultra-sound transducer operates at 40 kHz. The transducer is a piezo-electric material machined to such dimensions that it resonates at 40 kHz. When the oscillator is switched by the operator, it produces pulses of ultrasound, usually in the form of a code. The range of an ultrasound transmitter is usually limited to a few metres. One of the reasons for this is that the wavelength of the ultrasound is small (about 8mm in air). This is somewhat less than the dimensions of many of the objects in a room. Consequently, these objects cast sound shadows, which may interfere with the transmission of ultrasound to the receiver. The short wavelength also means that the sound radiates from the transmitter as a beam with an angle of divergence of about 20°. The receiver has a similar receiving angle. Therefore, the transmitter must be aimed fairly accurately at the receiver to ensure satisfactory control. Although ultrasound is a simple and inexpensive technique for remote control, these limitations make it less popular than other methods.

Infrared: In domestic systems, this is probably the most widely used of all control techniques. The transmitter consists of an infrared LED or maybe a bank of several of these, controlled by a pulse generator (see later). The receiver is a photodiode with its peak sensitivity in the infrared band. The signal picked up by the receiver is passed to a decoding circuit and, depending on the code group received one of a number of different functions are enabled. The transmitter produces a broad beam of infrared, which is reflected from the walls of the room. For this reason, it is not necessary to aim the transmitter accurately except at extreme range. Since the system operates in the infrared region of the spectrum, it is not affected by normal domestic lighting. Infrared is also transmitted by optical fibre, giving it an interference-free range of several tens of metres.

Cable: A direct wire connection means that both transmitter and receiver can be simple circuits, perhaps based on a pair of integrated circuits, a line driver and a line receiver. There is less chance of noise entering this kind of system than in any other system (except optical fibre). Cables can pick up electromagnetic interference, but the effects

MERLIN

Merlin is a network of radio telescopes centred on the Lovell telescope at the Nuffield Radio Astronomy Laboratories at Jodrell Bank, Cheshire. Its name is short for Multi-Element Radio-Linked Interferometer Network.

The principle of MERLIN is that, if we direct a number of widely spaced telescopes at a given object, and combine their signals electronically, the effect is as if we had used a single telescope with a very large reflector. The Merlin array, spread across the English Midlands, is equivalent to a reflector 230 km in diameter. Of course, most of the 'surface' of this dish is missing because the constituent reflectors occupy only a small fraction of the total area. Therefore, the signal obtained is weak. The advantage of the array is that it has extremely large *resolution*. It is able to pick out the finest details of the celestial object. The resolution of a reflector or lens depends upon the ratio between its diameter and the wavelength of the radiation. Radio waves are much longer than light waves, so radio telescopes normally have much lower resolution than optical telescopes. Increasing the apparent diameter of the dish gives Merlin a resolution equivalent to that of the Hubble Space Telescope.

The radio telescopes of Merlin are all remotely controlled from Jodrell Bank by signals sent at 9.6 KBD along landlines. These have about the same bandwidth as ordinary telephone lines. To keep the activity of the telescopes synchronised, timing signals are sent to each one from Jodrell bank along microwave links. Data returned from the telescopes is returned along 8 GHz microwave links. This data includes feedback from the local computers driving each telescope. The data is continuously displayed on a bank of monitors (Fig. 8.2).

The aiming of the telescopes must be precisely co-ordinated so it is not possible to operate the system with open loops. The feedback includes all current operating parameters of each telescope, such as its angles of elevation and azimuth. Also included in the feedback are pictures from several video cameras suitably placed at each site. All this feedback data is displayed on an array of monitors in the Jodrell Bank control room.

Figure 8.2 *These computers and TV monitors in the main control room at Jodrell Bank are at the heart of the MERLIN network. The most distant telescope of the network is located at Cambridge.*

There are inevitably times when communications break down. The programming of the computer at each remote telescope allows it to continue to track an object for up to ten minutes after a break in the control signals. After this time, the telescope is switched off and parked pointing upward to the zenith, to avoid damage from strong winds. This fail-safe approach is one that is an essential feature of many control systems.

of this can usually be minimised by using shielded cable. On the other hand, cabling over appreciable distances can be expensive. Newly built houses are being wired for remote control as well as other systems such as TV and Internet access. In older houses, one of the problems of remote control is that wiring is unsightly, often difficult, and therefore expensive to install. One of the ways of avoiding this is to use the existing mains wiring to provide the connections.

Home automation

The X-10® system is designed to allow householders to automate their homes by making use of the existing mains wiring. All connections to the system are made by plugging units into the mains wall sockets. There is a control box from which all other modules in other parts of the house are controlled. The other modules are of various types, mainly intended for switching domestic appliances such as table lamps, heaters, and fans.

When the control box sends a signal to a module, it first sends a *house code*. There are 16 possible house codes and their purpose is to identify signals produced in a certain house, so that they have no effect on devices in a neighbour's house. Next, the control box sends one of 16 *unit codes*, depending on which module is to be contacted. Then it sends a *control signal*. This may signal ON or OFF, to switch a device fully on or off. For certain appliances such as lamps, there are also DIM and BRIGHT commands.

It is possible to use a microcontroller to operate the control box automatically. The BASIC Stamp 2 microcontroller, for example, is provided with a number of commands to make it generate X-10 codes.

Radio: This operates over a greater distance than the other methods and there is no physical connection between transmitter and receiver. In many instances, it is not necessary for the transmitter to be aimed directly at the receiver. This makes it easy to control mobile devices ranging from flying model aeroplanes to Moon Buggies. In recent years, there has been increasing use of radio for short-range control. Part of the expansion is due to the availability of mass-produced transmitter and receiver modules operating on a fixed legal frequency (such as 418 MHz). By using these approved modules, manufacturers (and hobbyists) are able to design and produce low-cost radio control systems without the expense and effort of having them type-approved or licensed by the government (Fig. 8.3). These devices include radio controlled car immobilisers, central locking systems, doorbells, and garage door controllers. There is also an increasing range of inexpensive radio-controlled toys (see box, p. 196).

Figure 8.3 *The availability of radio frequency oscillator modules makes it possible for the home constructor to design and build a range of radio controlled systems. Here, a key fob contains the transmitter circuit for controlling a domestic security system from outside the house. The radio module is the light-coloured component at top centre of the board. The thick curving wire is the antenna.*

Bluetooth

Philips Semiconductors and their associates have developed a new technology for interconnecting the increasing number of electronic devices that are to be found around the home and office. The Bluetooth® system comprises digital semiconductor chips that incorporate radio frequency circuits operating in the 24 GHz band. It is expected that, in the first place, these will be used for short-range wireless communication between laptop computers and mobile telephones. They will also provide for wireless communication between a computer and a mouse. In the audio field, they will eliminate the need for cables connecting amplifiers to speakers. This system has obvious applications to remote control.

Keeping up?

1 What are the main stages in a remote control system?

2 Why is electronic feedback less often used in a remote control system?

3 What kinds of channel are used in remote control systems? What are their advantages and disadvantages?

Coding

A TV set or video controller makes use of infrared as the channel of communication. The controller generates a coded series of pulses. The code determines what happens when the transmission is received and decoded in the TV set. The pulses are not just bursts of infrared radiation but are modulated on to a high-frequency signal. A typical signal would have a frequency of 500 kHz. Fig. 8.4 shows what the modulated signal looks like. It consists of pulses of the 500 kHz carrier, all of equal length. The spaces between the pulses are of three different lengths. The shortest space represents digital 1. A space 50% longer represents 0. A space twice as long as the 0 represents a stop bit, S. In Fig. 8.4 the signal has five bits:

11010

On receiving such a signal, one of the output pins of the decoder goes high, perhaps changing the TV, or increasing the sound volume. A 5-bit code allows for up to 32 different commands. This system of sending signals by varying the timing between consecutive pulses is known as *pulse position modulation* (PPM).

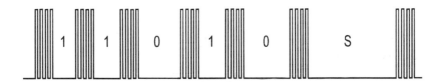

Figure 8.4 *In pulse position modulation, the high-frequency carrier signal is modulated by pulses. Spacing between pulses indicates logical bits (0 or 1) or the stop bit (S). The signal is repeated several times.*

The system has safeguards against incorrect action. One of these is that the decoder responds only if it receives the same code twice in succession. This avoids errors due to interference. Another safeguard is that the signal can be received only if the oscillator in the receiving circuit is running at the same frequency as that in the transmitting circuit. In other words, the receiver is tuned to a particular frequency and ignores signals sent at other frequencies. This allows other

Control systems at play – 1

The availability of inexpensive, type-approved radio modules has led to an increase in the range of radio controlled toys. Here, Emma and Laura discuss steering tactics for their radio controlled *Rising Bird* buggy. Laura holds the control unit, which has two joysticks for start/ stop, forward/reverse and left/right control.

systems such as controllers for the TV set, the video player and the audio system to be operated together in the same room. A similar arrangement is used with flying model aeroplanes. The frequency of the radio control transmitter is determined by a plug-in crystal. Enthusiasts gathering to fly their models can plug-in different crystals so that they are able to have several planes in the air simultaneously, without interference.

This scheme is not practicable for the dozens of controllers that may be found in a residential street today. When so many householders have garage door openers, car immobilisers, car security devices, and home security systems operating in a confined area, there is no way in which they can expect to agree to operate on particular frequencies. There is also the ever-growing problem that the radio spectrum is becoming increasingly over-subscribed. If there is any change in allocations, it is toward restricting the number of frequencies available for such users. As explained earlier, there is a trend to use type approved radio modules, all operating at the *same* frequency. The 418 MHz frequency is often used. The solution to communicating only with the intended device is to preface the control code with an ID code.

Fig. 8.4 shows a coder IC that produces signals suitable for a remote control system. The ID code is determined by the inputs to terminals *a* to *e* of the IC. Each terminal can be connected to 0V, the positive supply (+V) or left unconnected. This means that, instead of a binary input (0 or 1) we have a trinary input (0, 1 or X). With five trinary digits, the total number of possible IC codes is 3^5, which is 243 codes.

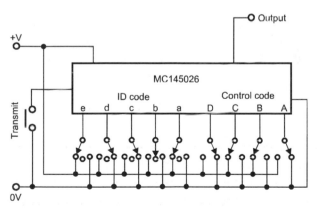

Figure 8.4 *An MC145026 coder IC, as used in the transmitter of a remote control system. If only a few ID codes are needed, the switches on inputs* a *to* e *may be replaced by wired connections to +V or 0 V, or they may be left unconnected.*

Control systems at play – 2

The Japanese company Keyence manufactures a pair of sophisticated Hi-Tech toys. One is the *Revolutor* H-610 radio controlled flying model helicopter. It is claimed that this is the smallest indoor helicopter in the World. It is 290 mm long, its main rotor is 350 mm in diameter, and it weighs only 100 g. It features a pair of piezo gyroscopes to give it stability in flight. It can fly free for up to 1 min, using battery power or for longer when connected to the controller by a light cable. It has a range of 10 m. It has a 4-channel proportional control system operating in the 40 MHz band.

The *Engager* GS III – E – 770 is described as a radio controlled saucer and has an unconventional appearance to support this description. It is 500 mm long and 400 mm wide. Like the *Revolutor*, it has on-board gyroscopes to maintain stability.

This is enough to cover most circumstances and avoid undue interference. The code transmitted for an IC switched as in Fig. 8.4 is 100X1. The MC145026 sends the codes as two long pulses for a '1', two short pulses for a '0' and a long pulse followed by a short pulse for an open circuit (X).

By switching codes, it is also possible to use a single transmitter to control up to 243 separate devices.

Following the ID code, the transmitter produces a four-bit control code, according to the inputs to terminals *A* to *D*. These are binary inputs and allow for up to 16 control codes. The decoder IC has a similar set of terminals. There are five ID code terminals, with three-way switches as in Fig. 8.4. The switches are set (or permanent wiring is used) to set the ID for the receiving device. When a signal is

ROVs

Remotely operated vehicles are used where there are regions to which it is too dangerous or too far for humans to go. In the oil and gas industries, underwater ROVs are widely used for surveys and inspection and for underwater repairs.

The Voyager® ROV, manufactured by Perry Tritech, is an unmanned ROV which can operate at a depth of 610 m. Special versions can operate down to 1500 m. The vehicle itself is approximately rectangular and is 107 cm long, 81 cm wide and 66 cm high. It is connected to the control console on the rig or boat by a multicore cable including optical fibre video channels, as well as electrical power and signal lines.

It is propelled by five DC brushless thrusters that propel the vehicle forward and in reverse, as well as from side to side. They also propel it up and down and rotate it. It is capable of speeds of 2.7 knots in the forward direction and 2 knots in reverse.

The Voyager has on board 1000 W of lighting for the video cameras and the still camera. Its complement of tools includes an electrically powered manipulator arm, rope and wire cutters, a scanning sonar, and a pipeline tracker system.

The Triton® ST is another ROV made by the same company. This is used for deepwater salvage and remote tool deployment. It has a maximum depth of 3000 m and is operated by a Pentium-based control system.

received, its ID code is compared with the setting of the ID on the receiver. If the two are identical a 'valid signal received' output goes high, and the control code appears on four *outputs* corresponding to A to D. This logic output is further decoded to put the appropriate action in hand.

If we are using such an IC to produce codes for a garage door opening control or a home security controller, there is no need for a 16-bit control code. All that is required is a single on-off code and this needs only one pulse to toggle the function on or off. In this case, all nine inputs of the coder can be used as ID inputs. This gives 3^9 ID codes, a total of 19683 codes. In practice, because of the way in which the codes are matched, the total is 13122. This gives a very low risk that two transmitters being used in the same locality will produce identical codes. This makes the IC very useful for security systems. On receiving and matching the 9-bit ID code, the 'valid signal received' output goes high. This output can be used to toggle the controlled circuit. For this kind of application some other ICs generate even more codes. For example, the HT651 has 18 trinary inputs, which give 3^{18}, or 387 *million* different addresses.

Test yourself

1 In most remote control systems, the feedback is:

> A negative.
> B non-existent.
> C visual.
> D proportional.

2 An ultra-sonic transmitter has a small range because:

> A of sound shadows.
> B ultra-sound has low volume.
> C humans can not hear it.
> D the receivers are insensitive.

3 Most remote control devices in the home use:

      A ultra-sound.
      B radio.
      C infrared
      D the mains wires.

4 For long-distance remote control, we use:

      A infrared.
      B radio.
      C fibre optics.
      D cable.

5 The number of possible 5-bit control codes is:

      A 5.
      B 32.
      C 387 million.
      D 243.

6 The X-10® system communicates by means of:

      A infrared.
      B X-rays.
      C radio.
      D the mains wires.

7 In PPM the codes are indicated by:

      A the length of the pulses.
      B the height of the pulses.
      C the timing between pulses.
      D the polarity of the pulses.

9
Robotics

According to *The Great Encyclopaedia of Universal Knowledge* (Odhams, c. 1948), a robot is:

> A machine that does all the work of a human being.

In spite of the vast development of solid state electronics and of computers since that time, most current definitions agree that the essential point about a robot is that it is a machine with behaviour that, to a greater or lesser degree, resembles that of a human. The term was originated in 1920 by the Czech playwright, Karel Capek. The title of his play, *R.U.R.,* was short for *Rossum's Universal Robots.*

The popular idea of a robot is a humanoid creature, such as C-3P0 in the film *Star Wars.* However, at least until such machines become more widely available for helping humans at work and play, a robot is more likely to look like the machine in Fig. 9.1.

On inspecting the robots that we see on the factory floor (for it is here that robots have made their greatest incursion), we note that each has a sufficient but restricted range of human-like abilities. Those tasks that they perform they perform reliably and well, often with greater precision and power than the humans they replace. They have greater endurance of hardship too, and never need a tea break. We may find robots working steadily in environments where it is not safe or acceptable for humans to go.

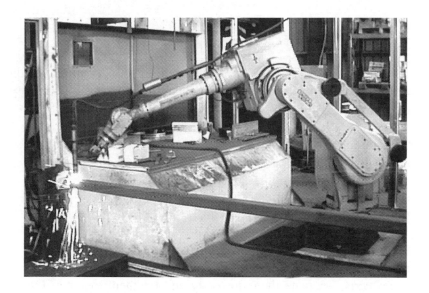

Figure 9.1 *An articulated arm robot reaches out to punch a hole in one end of a steel girder.*

Given suitable training or experience, most humans can do most things reasonably well. There are experts in many fields but most of us can wield a hammer, fix a tap washer, or ride a bicycle if we are taught how. This is because all humans have the same skeleton with the same joints and muscles. Mechanically, we are all very similar. We all have the same kind of control (= nervous) system, able to control our bodies. However, most robots are able to perform only a limited range of tasks and therefore they exist in many different forms.

The two main forms of robot are:

- Manipulators (robot arms).
- Moving platform robots (rovers).

Manipulators

These are primarily used on the production lines of factories, for the purpose of speeding assembly and in general increasing productivity. These reasons for this are:

They can move faster than human operators can.
They make fewer mistakes.
They do not become tired.
They can work a 24-hour day.
They can perform dangerous tasks and work in hazardous areas.

Within the group of manipulators there are subdivisions depending on what they manipulate. Some are used for assembly. They are able to pick up parts, orientate them correctly and fix them together. Others are used for fatiguing but skilled work such as welding. Another common task for robots is paint spraying. With flexible arms, they can direct the spray into awkward corners and, if carefully programmed, do not accidentally leave essential areas unsprayed. Often the same type of arm can be used for a variety of tasks. It just needs fitting with an appropriate gripper, a welding tool or a paint spray. It may also need appropriate sensors to provide feedback.

One of the more important features of a manipulating robot is the number of ways in which its arm can be moved. Basically, there are two kinds of movement:

- Rotation – one part rotates with respect to another.

- Translation – one part slides with respect to another.

Fig. 9.2 shows an *articulated arm* robot with rotation about six axes:

- θ is the horizontal rotation of the first arm member about the base.

- ω is the vertical angular motion of the first arm member about the base.

- U is the vertical angular motion of the second arm member with

Figure 9.2 *This articulated arm robot, a Hitachi MR6100, rotates around six axes, as shown by the arrows.*

respect to the first arm member.

• α, β, and γ are rotations at the 'wrist' in three mutually perpendicular axes.

Some models have only two axes of rotation at the 'wrist'. In the *polar* configuration, two of the joints rotate and the arm slides in and out to produce the third axis of movement. In the *Cartesian* configuration, the arm slides in three perpendicular directions. This is closely related to the movement of *gantry robots*, which have the same three-axis movements but are supported on a rigid framework (gantry) to allow them to handle heavier loads (Fig. 9.3). The movements of the robot illustrated in the figure are computer controlled. It is programmed to receive the blocks of pages that have been banded together by another robot, and to place them in stacks of 12 blocks. The exact position for stacking each block in each layer is programmed into the computer as

Figure 9.3 *This gantry robot (1) uses a gripper (2,3) specially designed for handling banded blocks of printed paper. Here the final block (4) is being added to a stack (5) of twelve blocks. When the stack is complete, it will be taken away to the bindery. The gantry is very robust, as can be seen from the dimensions of its supporting legs (6).*

a set of three co-ordinates. When two stacks are complete, the computer signals to the operator that they should be taken away to the next stage of production.

Programming a Cartesian or gantry robot is relatively simple because it is easy to express the required gripper positions in terms of perpendicular X, Y and Z axes. Cylindrical robots are also easy to program because there is only one axis of rotation. The most difficult to program are the articulated arm robots, with their six axes. Their

Vital statistics

The F series of articulated arms manufactured by Kawasaki is representative of robots commonly found on the factory floor. The F series has 6 or 7 axes of rotation and the arm usually ends in a gripper. The robot may be stationary, mounted on the floor or ceiling, or it may have limited but useful mobility on a tranversing unit.

The F series is intended for material handling, welding, sealant dispensing, washing and clean room operation. Robots used in a clean room can be sealed so as not to contaminate the working area. When it is essential for an operation, say the assembly of parts of a satellite, to be completely free from dust and microbial contamination, a robot has many advantages over a human worker.

One of the lighter models in the F series is the FS45N, which is capable of lifting a load of up to 45 kg. The arm has a reach of 1.8 m, yet its repeatability (the accuracy with which the gripper can be positioned) is as small as ±0.l5 mm. The slightly lighter model, the FS06N has a repeatability of only ±0.05 mm.

The arms of the F series can be run for 10 000 hours before requiring routine maintenance. They need an overhaul every 40 000 hours. Their endurance far exceeds those of humans.

Toward the other end of the range is the heavy-duty ZX200S arm. This can raise a load of up to 200 kg. It has a horizontal reach of 2.651 m and a vertical reach of 3.415 m. Even a large arm such as this is able to position itself to within 0.3 mm.

many axes allow great flexibility of positioning so that they are particularly suitable for tasks such as paint spraying.

Keeping up?

1 Define a robot.

2 What are the two main kinds of robot?

3 What are the advantages of robot manipulators over a human worker?

4 Describe a robot arm and its axes of motion. What are the main configurations?

Controlling manipulators

In principle, the control systems of robots are very similar to the systems already described in this book. Systems may involve feedback loops and may include servo systems. The systems may be controlled by logic circuits, or more often, by programmed microcontrollers or computers. In the more advanced robots, programming may include fuzzy logic and neural networks.

On a production line, each robot is usually autonomous. It has its own individual control system. However, like human workers, robots must cooperate. This is usually achieved by a kind of 'handshaking'. When robot A has finished its task, it signals this fact to robot B, the next robot along the line. This signal may be no more complex than a logical high output fed to an input of robot B. On receiving this input, robot B will send a 'busy' signal to A. On receiving this, A will temporarily stop its operations. When B has completed its present task and has managed to pass its completed workpiece to robot C further along the line, it will signal back to A that it is now ready to receive new work. A picks up the piece, or in some other way passes it to B, and then turns to receive a new workpiece from the input stack. In this way, the workpieces are passed along the line from robot to robot.

One class of programming that is commonly used with robots is a version of the expert system. Either the rules of behaviour are programmed directly into the robot's memory (Fig. 9.4) or they are acquired by learning. To learn how to perform a given task, such as spray-painting an automobile door, the robot its first switched to learning mode. Then an expert, using manual controls takes the robot through the various stages in the painting routine. At every instant, the movements and action are recorded in the robot's memory. The robot has learned the task. It can then 'replay' the data stored in its memory and repeat the actions of the expert exactly. This feature gives robots

Figure 9.4 *This is the control panel of the robot pictured in Fig. 9.2. Below it hangs the teaching pendant. This is a hand-held programmer used by the expert for teaching the robot what to do. A programming session may take a few hours but, after that, the robot repeats the expert's actions exactly every time it is run.*

a degree of flexibility that is important economically. During the course of its 'life', a robot can be trained to perform a wide range of different tasks. In this way it is superior to a machine dedicated to just one task and therefore redundant when that task has become obsolete.

Like other control systems, a robot has sensors linked by a control system to actuators. It is the special sensors and actuators that make the robot's control systems of special interest.

Driving manipulators

The earliest manipulators were powered by hydraulic or pneumatic drives. These are still in use, hydraulic power often being used where heavy loads are to be moved. Hydraulic systems rely on a liquid such as oil. The system requires a pump, usually electrically driven, to provide the necessary oil pressure. The oil passes through a valve, which is often controlled electrically by a PLC or computer, to drive an actuator, often a piston and cylinder (Fig. 9.5). A pneumatic system is similar, using compressed air as the fluid.

It might be thought that, because the control units are electrical or electronic, it would be best to use electrical actuators to drive the robot. While this is true in some instances, it is often preferable to use a hydraulic or pneumatic system, regulated by electrically operated valves. These are referred to as electro-hydraulic and electro-pneumatic systems. This distinguishes them from the purely hydraulic or pneumatic systems that are used for the control of much industrial plant. Below we compare the effectiveness of electrical, electro-hydraulic and electro-pneumatic systems as they apply to the control of robots.

Direct use of electrical power has the advantage that the electric motor or solenoid is simply driven by the relay or transistor output from a

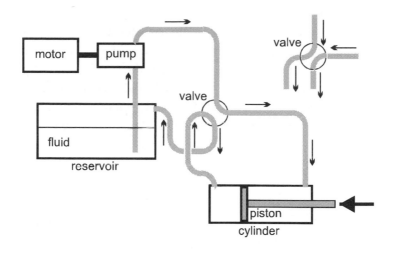

Figure 9.5 *A hydraulic system drives a linear actuator, consisting of a piston and cylinder. When the electrically controlled valve is rotated by a quarter of a turn (see inset), the pressure is applied to the other side of the piston, which moves in the opposite direction.*

controller. The actuator is usually an electric motor. An electric motor produces a continuous rotary motion with a high rate of revolution. This is seldom required in fixed robots of the manipulator type, in which angular motion of the arm is usually limited to 300° or less. Usually, the motor drives the arm through reduction gearing. This has the advantage that it allows a relatively small and light motor to be used to lift a heavy load, or to exert strong force. However, the gearing often makes the action slow. It is linear motion that is more frequently needed in a manipulator. Even the action of bending or straightening the arm is better done by a linear actuator than by a rotary one, putting the electric motor at a disadvantage. A solenoid produces linear motion directly, but the range is usually limited to a few millimetres. In addition, there is the problem that electric motors and solenoids become very hot if run continuously in an enclosed space.

Much higher power can be obtained from a hydraulic system. The source of energy is the hydraulic pump, usually driven by an electric motor. The prime advantage of hydraulics is that it provides a powerful drive to linear actuators in the form of a cylinder and piston. There may be one-way action in which the piston is moved by hydraulic pressure on one side of it. It moves against the action of a spring, which restores the piston to its former position when the pressure is released. There are also two-way actuators (Fig. 9.5), in which hydraulic power is used to push the piston from either side. Both types of actuator are powerful and fast moving, with a long travel. They cater for most of the driving requirements of a manipulator robot. The ultimate control of an electro-hydraulic system is electrical, by means of the solenoid-operated valves that admit the hydraulic fluid to the cylinders.

Rotary actuators are also used in hydraulic systems (Fig. 9.6). The cylinder has a rotating vane that is turned clockwise or counter-clockwise by admitting fluid through one or the other of the two ports. The rotary action is restricted to about 300° but, as explained above, this is as much as most robots require.

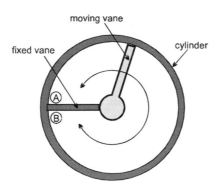

Figure 9.6 *The rotary hydraulic actuator has two input ports, A and B. The moving vane turns clockwise when fluid is admitted through port A and is exhausted through port B. The action is reversed for counter-clockwise rotation.*

Hydraulic systems are often used in plant that is handling flammable materials. An electrically powered device, or the electric cable supplying the current to such a device, may occasionally fail. The result may be excessive heat leading to ignition of flammable fumes. In an electro-hydraulic system, the solenoid valves are located at a distance from the processing area, often in a separate fireproof enclosure. Hydraulic lines run from there to the robots. Thus, there is no electrical equipment or wiring in the area of fire risk.

Pneumatic systems operate in a way similar to that of hydraulic systems and are used to drive linear and rotary activators. Pneumatic systems can also drive air motors. Usually pneumatic systems are faster acting than hydraulic systems but, size-for-size, the actuators are less powerful than their hydraulic equivalents. A second disadvantage arises from the compressibility of air. This introduces an element of imprecision into the system. By contrast, a hydraulic system uses a liquid, and all liquids are incompressible. There is a direct transfer of pressure from the compressor pump to the actuator.

Keeping up?

5 Describe how a robot may be taught to perform a given task.

6 Compare the features of electrical, electro-hydraulic and electro-pneumatic actuators.

Sensors for manipulators

A robot has to know what is going on around it. If it is of the common type known as a *pick-and-place* robot, so suited to assembly lines, it has to find the object that is has to pick up, and has to be able to place it in exactly the right location. Sensors are very necessary for this type of operation.

A robot may have several dozen proximity sensors to detect when its arm has arrived at the correct point in space, or when the workpiece is present ready to be picked up. A robot that performs a repetitive task may stack its product until an optical sensor tells it that the pile is high enough to be passed on to the next stage of production. Often the optical sensor consists of no more than a photodiode receiving a beam of infrared radiation. When the stack is high enough, the beam is blocked and a signal goes to the control system. The robot takes action accordingly.

Optical sensors, magnetic sensors, microswitches and many other devices are employed as proximity sensors. Occasionally a less obvious technique is used. Normally a manipulator holds the object it is working on by clamping it in a gripper at the end of the arm. For handling light parts of complex shape (for example, plastic mouldings) a robot may be equipped with a suction handler. This consists of a structure shaped so as to make contact with the moulding at certain points. Contact is made at all points only if the moulding is orientated correctly, and is therefore ready for processing. There are suction pads at the contact points and these hold the moulding securely in place while it is being processed. Thus, the object is retained by suction instead of a mechanical gripper. A pressure sensor measures the air

Social consequences

One of the obvious consequences of using robots in factories is that fewer workers are required. The social effects of this are many, and may be difficult to resolve. As far as the factory itself is concerned, one of the more noticeable effects is that the factory no longer needs a large canteen to feed its staff on site. Robots do not need snacks and meals. The canteen is closed down and the few remaining human staff eat in local cafes.

pressure in the suction pipes. If the moulding is in place and correctly orientated, the system is practically sealed and the pressure falls sharply. This fall is detected by the pressure sensor, which informs the control system that the moulding is ready for processing. However, if there is no moulding present, or if it is not correctly orientated, there is a leakage at the suction points. Pressure does not fall and the control system becomes aware that something is wrong. It may then sound an alarm to call an operator to investigate.

Tactile sensors are another category of importance in manipulators. They are usually placed on the gripper so that the shape, size or weight of the object may be sensed. Levers with microswitches attached are among the simplest of these sensors. It is also possible to use strain gauges, and piezo-electric sensors. In the latter case, a wafer of quartz or of lithium niobate is fixed to a face of the gripper. As the gripper grasps the object, the force of gripping distorts the crystalline structure of the sensor. This causes an e.m.f. to be generated, and this is detected and analysed by the control system.

The use of tactile sensors as backpressure sensors is important in robots. A gripper can exert considerable force, which may be essential when holding a robust, massive object. Unfortunately, the gripper could easily crush a more delicate object. By programming the robot to respond to backpressure, the robot can be prevented from damaging the object it is handling. Robot arms can handle eggshells without cracking them. For this reason, robots are often used to handle delicate objects. They are less likely than their human counterparts to drop or smash them.

A special type of tactile sensor is often used to guide a robot when it is welding. A sensor probe is attached to the gas nozzle of the welding head. When the sensor is deflected by contact with the workpiece, magnetic bodies at its base cause variations in the magnetic field threading a number of piezo-magnetic resistors. The variations in current produced in this way are analysed and interpreted. Initially, the robot is taught the approximate path it is to follow. Then, with the sensor attached to the welding torch, the torch is guided along the welding path with the sensor detecting the actual location of the weld seam. Signals from the sensor are used to correct the roughly programmed path, so that the welding torch is directed more precisely at the seam as the corrected path is learned.

In some instances, it is necessary to detect an object without actually making contact with it. Capacitative and inductive sensors are specially suited for this purpose, the latter being particularly suited to detecting the proximity of metal parts. When a robot is tooled for welding, it needs a sensor to detect the edges of the workpiece. This may be a laser sensor. A concentrated laser beam is directed at the workpiece and is reflected on to an array of photodiodes. The pattern of reflection detected by the photodiodes is analysed by the control system, which then calculates the position of the workpiece.

Keeping up?

7 List some of the sensors used by robots, and explain how they are used.

Moving platform robots

The popular view of a robot is not a manipulator operating from a fixed base, but a mobile robot, with a decidedly human appearance. There are indeed robots of this type but at present they mainly appear in fictional film dramas, or as advertising gimmicks at exhibitions. For general mobility, it is easier from the viewpoint of construction and control for the robot to move on wheels. Usually the base of the robot has three wheels at the vertices of an equilateral triangle, or four wheels at the corners of a square. These arrangements give greatest stability. Various configurations of steering and drive wheels are in use. On three-wheeled platforms, it is most usual for there to be one steering wheel and two drive wheels. On four-wheeled platforms, there may be two wheels for driving and two for steering. However, another workable system is for the two front wheels to be used for both steering and driving, with separate motors. The two rear wheels are then simply castors.

In spite of the simplicity of using wheels, there has been much research into other forms of propulsion. One of the problems is that wheels can not negotiate steps and are often useless on rough surfaces unless they are large. Legs have the advantage here, but introduce the complication that the robot must remain balanced when it lifts one leg (or more) to move it forward. Humans achieve this with only a pair of legs because they have balance sensors in the inner ear. To program a robot to perform the same action is making things more difficult than they need be.

Some very successful robots have been built using six legs. Three legs (the front and rear legs on one side and the middle leg on the other side) are lifted and moved forward together. The other three legs remain on the ground and make a firm triangular base to give stability. Another successful approach for use on rough ground is to use caterpillar tracks. These are only a few of the ways, many of them very ingenious, that have been devised to give the robot its mobility.

Associated with mobility are the problems of avoiding obstacles and the more positive problem of steering toward a known target. Many robots are equipped with tactile sensors that touch objects ahead. Often the sensor is a long, thin, stiff wire with a microswitch at its base. When the switch is closed the robot reverses its motion for a short distance, turns left or right at a slight angle and then resumes in a forward direction. Robots with these facilities have been designed to learn their way through a simple maze. Interesting though these may be, they are really only 'toy' applications.

A well-tried technique for steering a mobile robot is to use two light sensors to detect a 'path' painted on the floor. A beam from a lamp mounted on the underside of the robot is directed down on to the painted stripe (Fig. 9.7). The reflected light is detected by two photodiodes or phototransistors, marked A and B in the figure. Assuming that the strip is white and the floor is grey, the robot steers straight ahead when 'grey' is detected by both sensors. When the robot

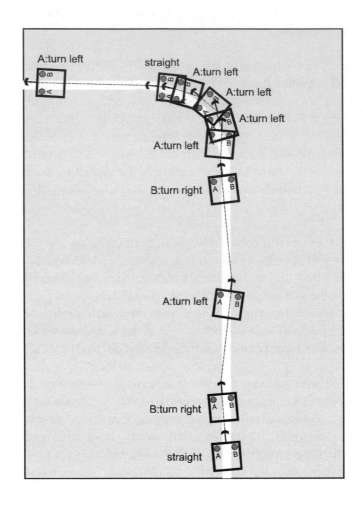

Figure 9.7 *A mobile robot with two optical sensors corrects itself when it strays from the white painted path and successfully negotiates a corner.*

begins to stray from the path, or when it comes to a corner, one of the sensors will receive more light than usual, reflected from the white stripe. The control system responds to this by altering the steering to bring both sensors to the 'low light' state again.

A similar system is based on detecting an electromagnetic field. The path is created by burying a wire in the floor and energising it with a high frequency electric current. The magnetic field is detected by two magnetic sensors mounted on either side of the robot. The steering mechanism is adjusted to keep the magnetic intensity equal for both sensors.

This kind of self-steering robot has many applications. In factories, such a robot can bring supplies of parts to workstations at various points along a production line. The robot can be programmed to go to the supply centre, pick up a prescribed load of parts and convey them to the workstation where they are needed. They may be used for carrying hazardous loads or in hazardous areas. In a hospital, they may be used for transporting drugs to the wards and, unlike a human courier, are proof against attack by would-be thieves.

Navigating by painted tracks or buried cables is common practice with relatively simple robots. More sophisticated robots are able to find their way about a building without these aids. They may have tactile or optical devices to sense the features of their immediate surrounds, but their most informative feature is a complete map of the building stored in memory. At any instant, the robot 'knows' where it is and what route to take to get to another part of the building. Some robots are able to update their internal map as a result of experience. For instance if its sensors detect that its way is blocked by an obstacle, it waits a while for the obstacle to move (it might be a pair of people standing talking in the corridor). If the obstacle moves away after a while, it is taken to be an impermanent feature and is 'forgotten'. If, on the other hand, the obstacle does not move away it is taken to be a new feature of the environment. Its presence is noted on the internal map and the robot works out a way to pass it by.

Programs that give the robot the mapping features described above are highly complex and may often involve fuzzy logic and neural networks. Programs of similar complexity may also be used to give the robot the ability to visually recognise different kinds of object. This has practical applications in 'pick and place' robots where the robot may be required to recognise and pick up one of a number of different objects (such as parts of a compact disc player) placed haphazardly on a conveyor belt.

The robot 'eye' is a video camera producing a two-dimensional image of the object. This is in the form of rows and columns of pixels, which will be either white (the colour of the background) or grey to black (the object). It is fairly easy to devise software to analyse the image. A number of different parameters may be obtained:

- Area: more or less proportional to the number of grey/black pixels. This gives a measure of the size of the object.

- Perimeter: how many grey/black pixels have a white pixel adjacent. This estimates the irregularity of the outline of the object.

- Aspect ratio: maximum dimension divided by minimum dimension. This differentiates between rounded objects and elongated objects.

In practice, the way these measurements are taken may need to be refined but the principle is clear. One way of tackling the programming is to devise a series of IF ... THEN statements to sort the images and give a name to them. It is possible to analyse the image in this way to identify the object (from a limited number of possibilities) and to ascertain its orientation prior to picking it up. Another approach is to train a neural network (see Chapter 7). Using supervised training, the robot is presented with an assortment of objects, which it examines and analyses to find its parameters. As each object is presented, the robot is told which kind it is. It builds up a table of typical parameter values for each kind of object. Then, it is told to build a neural network based on this data. After this, it is able to distinguish between the different kinds of object.

Test yourself

1 The first person to use the word 'robot' was

       A Bill Gates.

       B Isaac Azimov.

       C Karel Capek.

       D H.G.Wells.

2 A robot of the Cartesian configuration has:

       A 3 axcs.

       B 4 axes.

       C 5 axes.

       D 6 axes.

3 A robot with six axes of rotation is called an:

       A cylindrical robot.

       B gantry robot.

       C articulated arm robot.

       D polar robot.

4 The most powerful type of robot actuator is:

       A electro-pneumatic.

       B an electric motor.

       C a solenoid.

       D electro-hydraulic.

5 A tactile sensor responds to:

       A light.

       B touch.

       C temperature.

       D pressure.

Appendix

Answers to numerical and multiple-choice questions

Chapter 1

Keeping up?   5. +3.5 V.

Test yourself   1 D   2 B   3 A   4 B   5 A
                         6 A   7 D   8 D   9 A   10 B

Chapter 2

Test yourself   1 D   2 B   3 A   4 D   5 C
                         6 B   7 C   8 B   9 A   10 C

Chapter 3

Test yourself   1 C   2 C   3 A   4 C   5 D
                         6 B   7 C   8 A   9 B   10 D

Chapter 4

Keeping up?   5. See Fig. A.1.

Test yourself   1 C   2 A   3 D   4 B   5 B   6 D   7 B

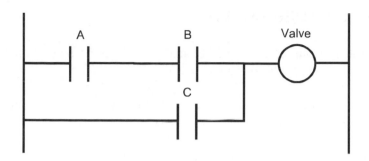

Figure A.1 *The answer to Question 5.*

Chapter 5

Test yourself   1 B   2 A   3 C   4 B   5 A
                            6 B   7 B   8 C   9 B   10 A

Chapter 6

Test yourself   1 D   2 D   3 B   4 C   5 A
                            6 B   7 C   8 C   9 D   10 A

Chapter 7

Test yourself   1 B   2 D   3 B   4 C   5 A
                            6 C   7 D   8 A   9 C   10 C

Chapter 8

Test yourself   1 C   2 A   3 C   4 B   5 B   6 D   7 C

Chapter 9

Test yourself   1 C   2 A   3 C   4 D   5 B

Acknowledgements

The author thanks the companies and organisations listed below for permission to take photographs on their sites:

Cathay Pacific Airways, Perth International Airport, Western Australia (pp. 3 and 80).
Eastern Generation Ltd., Ironbridge, Shropshire (p. 52).
Glanbia Foods Ltd., Oswestry, Shropshire (p. 73).
Jarrold Printing Ltd., Norwich (pp. 108 and 206).
Nuffield Radio Astronomy Laboratories, Jodrell Bank, Cheshire (pp. 121 and 191).
Bossong Engineering Pty. Ltd, Canning Vale, Western Australia (pp. 203, 205, and 209).

The author also wishes to thank the following individuals for helpful advice and assistance in explaining how they make use of electronic control systems:

Malcolm Ackerley (Rhône-Poulenc Ltd., Norwich), George Aitchinson (Manufacturing Engineer, Siemens Automotive Systems Ltd., Telford), R. Byrnes (Richard Burbidge Decorative Timber Ltd., Oswestry, Shropshire), Russ Cason (Senior Electrical Engineer, Kronospan Ltd.), Roger Doyle (Engineering Director, Machines and Systems (Design) Ltd., Wolverhampton), Alwin Hughes (Glanbia Foods Ltd.), Phil Ludgate (Parts Engineering Manager, Ricoh UK Products Ltd., Telford), Carl Maxwell (Principal Electronic Systems Engineer, Lucas Aerospace, Wolverhampton), Ian Morison (NRAL, Jodrell Bank), Colin Myers (Engineering Manager, Cathay Pacific Ltd.), Dave Potter (Head of Process Control Section, Eastern Generation Ltd.), Philip Ringwood (Electronics Engineer, Jarrold Printing Ltd.), Clayton Steele (Automation Technical Manager, Bossong Engineering Pty. Ltd.), Nick Thompson (Kronospan Ltd, Chirk), Chris Whitley (Principal Electronic Systems Engineer, Lucas Aerospace, Wolverhampton).

Software

The following software has been used in preparing some of the illustrations in this book:

Easy NN is a PC program written by Stephen Wolstenholme and is described as 'The easy way to build neural networks'. A restricted version is available for downloading as shareware from: http://www.tropheus.demon.co.uk/

Electronics Workbench® EDA is published by Interactive Image Technologies Ltd, Toronto, Ontario, Canada. http://www.interactiv.com/

FUZZLE™ is produced by Modico Inc., PO Box 8485, Knoxville, TN 37996-0002, USA. http://www.modico.com/

fuzzyTECH™ is produced by Inform, Pascalstrasse 23 D-5 1000, Aachen, Germany. http://www.fuzzytech.com/

SPLat/PC is the language for programming SPLat PLCs, manufactured by SPLat Controls Pty. Ltd., Seaford, Vic, Australia. http://www.microconsultants.com/

Index